Tobias Wirth

Functional Methods for the Solution of One-Dimensional Quantum Systems

Tobias Wirth

Functional Methods for the Solution of One-Dimensional Quantum Systems

Fusion Hierarchy, Functional Bethe Ansatz and Non-Linear Integral Equations

Südwestdeutscher Verlag für Hochschulschriften

Imprint
Any brand names and product names mentioned in this book are subject to trademark, brand or patent protection and are trademarks or registered trademarks of their respective holders. The use of brand names, product names, common names, trade names, product descriptions etc. even without a particular marking in this work is in no way to be construed to mean that such names may be regarded as unrestricted in respect of trademark and brand protection legislation and could thus be used by anyone.

Publisher:
Südwestdeutscher Verlag für Hochschulschriften
is a trademark of
Dodo Books Indian Ocean Ltd., member of the OmniScriptum S.R.L Publishing group
str. A.Russo 15, of. 61, Chisinau-2068, Republic of Moldova Europe
Printed at: see last page
ISBN: 978-3-8381-2387-5

Zugl. / Approved by: Hannover, Leibniz Universität, Diss., 2010

Copyright © Tobias Wirth
Copyright © 2011 Dodo Books Indian Ocean Ltd., member of the OmniScriptum S.R.L Publishing group

Contents

1	**Introduction**	**3**
	1.1 Motivation	3
	1.2 Brief Historic Overview and Outline	4
2	**Algebraic Principles**	**9**
	2.1 Quantum Inverse Scattering Method	9
	2.2 Algebraic Bethe Ansatz	17
	2.3 Fusion in Auxiliary Space	18
3	**Scalar Products for the Diagonal XXZ Spin Chain**	**23**
	3.1 Auxiliary Function	24
	3.2 Integral Representation for the Determinant Formula	29
	3.3 Generating Function of the Magnetization	32
4	**Separation of Variables**	**37**
	4.1 Functional Bethe Ansatz	38
	4.2 TQ-Equation	46
	4.3 Spin Boson Model	49
5	**XXZ Model with Twisted Boundary Conditions**	**57**
	5.1 Baxter's Method and Separation of Variables	58
	5.2 Fusion Hierarchy and Truncation Identity at Roots of Unity	60
6	**Non-linear Integral Equations for the XXX Spin Chain**	**63**
	6.1 Single Non-linear Integral Equation	63
	6.2 Y-System and Non-linear Integral Equations	67
7	**Conclusion and Outlook**	**77**

CONTENTS

Appendices		81
A	Density Function for Diagonal Boundaries	83
B	Comparison to Numerics for Diagonal Boundaries	85
C	Similarity Transformations for Truncation Identity at Roots of Unity	91
D	Equivalent Form of Fusion Hierarchy	93
E	Calculation of Driving Terms	95
F	Contributions from the Lattice to the Eigenvalue	99
Bibliography		103
Publications		111
Acknowledgements		113

Chapter 1

Introduction

1.1 Motivation

Integrable spin chains, just as all exactly solvable models, provide one of the very few possibilities to study many body physical systems as a whole, i.e. without approximations or the need to apply perturbation theory. The methods in this field are developed quite far and current research is concerned with the calculation of physical observables from correlation functions and the determination of the spectrum for chains with non-trivial boundary conditions. All known exactly solvable integrable models, including the spin chains, are one-dimensional restricting their direct applicability to experimental situations. However they always serve as a starting and reference point for treating strong correlated systems with quantum fluctuations. Furthermore the possibility remains to draw conclusions from the low dimensional studies to situations of higher dimensional models e.g. by coupling several one-dimensional systems. In the course of time the progress in material science led to the discovery and realization of more and more quasi one-dimensional systems in solid state physics. Nowadays excellent realisations of spin chains are known to be $KCuF_3$, Sr_2CuO_3, Cs_2CuCl_4 and CuPzN [29]. On the other hand the experimental techniques are advancing in a rapid pace and signatures of one-dimensional physics can be directly observed (e.g. [20, 37, 50, 92]). Another, at first sight, purely theoretical application of exact solutions is that results provide excellent benchmarks for numerical approaches like the density matrix renormalization group, see e.g. [16] where transport phenomena for a lattice model are studied using DMRG and the boundary sine-Gordon model as the continuum limit of the original model. These numerical approaches may then be applicable to more realistic models and higher dimensions.

However the impact is not only restricted to the realm of solid state physics. In the field of quantum optics the advances in producing optical lattices provide the setting and tools to experimentally construct low dimensional systems. In this context the thermal energies reachable by cooling are still above the magnitude of the energy of spin-spin interactions and hence a realisation of spin systems in optical lattices is subject of future experiments [48]. A different application is to use spin chains as a channel for short distance quantum communications which was proposed in [15] and attracted considerable interest in the community of quantum information. Furthermore, string theorist were able to relate anomalous scaling dimensions to the spectrum of integrable spin chains in the context of AdS/CFT correspondence thus opening up a new realm of mathematical applications [81].

In the next section we will summarize the historic developments of integrable models with emphasis on spin chains and the present situation which in turn will lead to an outline of this thesis.

1.2 Brief Historic Overview and Outline

In the 1920's Werner Heisenberg [39] and Paul Dirac [25] worked on the question of how to describe ferromagnetism as a phenomenon of spin interactions. The established Weiss' theory was only formally satisfactory as it was based on the assumption that an aligning force on an atom arises from all other atoms in the crystal but the origin of this force was unknown. Heisenberg's fundamental idea was to combine the Coulomb interaction, which alone is non-magnetic, with Pauli's exclusion principle. The combination turned out to be sufficient to reproduce the results of Weiss' theory of magnetism [42, 79]. To prove this he used the Coulomb exchange interaction introduced by himself earlier [40] which led to an interaction energy between pairs of electrons given by the so-called exchange integral. In 1929 Dirac derived an explicit expression for a hamiltonian and Heisenberg was able to derive Weiss' equations under the assumptions that only nearest neighbor interactions are relevant, that the exchange integrals are the same for all electron pairs and that the exchange energies exhibit a Gaussian distribution [42, 79]. These investigations led to the model now known as the Heisenberg model where relevant contributions arise only from nearest neighboring spins and the exchange energy J is the same for all pairs of spins

$$\mathcal{H}_{\text{Heisenberg}} = -2J \sum_{\text{neighbors}} \mathbf{S}_i \cdot \mathbf{S}_j \,.$$

This model laid the ground for a broad field in solid state and mathematical physics and models descending from this very basic idea are still in consideration of ongoing work.

The Heisenberg model itself is formulated in three space dimensions but the vastest development over the years since introduction was made for the simplest case of one-dimensional chains. As worked out several years later the reason for the success of exact

1.2. Brief Historic Overview and Outline

analytical methods in one dimension is the underlying algebraic structure assuring *integrability* of certain models.

Even the most simple case of the Heisenberg model where the anisotropy of the crystal structure permits the spins only to be orientated along the z-direction is not solved on a three-dimensional lattice. However aligning the spins on a chain allows for an exact solution of the partition function and further physical properties [9]. The case just described is the Ising model already introduced in 1925 [44].

The Heisenberg model in one dimension was first solved by Hans Bethe [13] using a method now known as the coordinate Bethe ansatz. Although in his paper from 1931 he was confident to extent his method to lattices of higher spacial dimension[1], this generalization was not possible. Nevertheless his work is still one of the most cited in this field and his results could be included in a more general framework for one-dimensional models around 1978-79. Today this framework utilizing the *Yang-Baxter algebra* is summarized by the names of Quantum Inverse Scattering Method and the algebraic Bethe ansatz (if applicable). The pioneers of this machinery, the *Leningrad group* around Faddeev [30], were influenced by or found striking similarity to the so-called transfer matrix method of Baxter on two-dimensional lattice models of classical statistical mechanics [9]. In both methods the hamiltonian is generated from a member of a commuting family of operators called transfer matrices yielding other integrals of motion as well. Furthermore an unexpected connection to the factorized scattering method in quantum N-body systems of one-dimensional particles developed by C.N. Yang [98], McGuire [67], Berezin *et al.* [12] and Brezin and Zinn-Justin [17] leads to a nice graphical interpretation of the formalism. Later on an abstract algebraic point of view was constructed by Drinfeld [26] introducing *quantum groups* as deformations of universal enveloping algebras of semi-simple classical Lie algebras and interpreting them as a special class of Hopf algebras. The studies in this thesis rely on the Quantum Inverse Scattering Method and its algebraic background concerning the Yang-Baxter algebra and its mathematical details will be reviewed in chapter 2. The power of this method was expressed by the number of models being embedded into its realm e.g. quantum nonlinear Schrödinger equation, Toda lattice, sine-Gordon model [32], one-dimensional Hubbard model and particularly its limits to various spin chains [28]. Although not all of them can be solved using the standard Bethe ansatz procedures, the algebraic background provides the starting point for the analysis of the system.

Today interesting models include, among others, spin chains with non-trivial boundary conditions, i.e. boundary fields at each end of the chain. Physically these systems describe open spin chains present in certain transitional metal compounds like Sr_2CuO_3 [4]. A one-dimensional chain inside a macroscopic material can be achieved by anisotropic coupling

---

[1] "In einer folgenden Arbeit soll die Methode auf räumliche Gitter ausgedehnt und die physikalischen Konsequenzen bezüglich Kohäsion, Ferromagnetismus und Leitfähigkeit gezogen werden." H. Bethe, 1931, [13]

1. Introduction

of atoms such that interactions in some spatial dimensions are irrelevant compared to the distinguished direction forming the chain. These chains are then broken apart by impurities present in the material which may also give rise to an effective boundary fields on the ends. The inclusion of boundary conditions in the Quantum Inverse Scattering Method was achieved by Sklyanin in 1987-88 [83] and opened up new possibilities of studies.

In this thesis we will consider the spin-$\frac{1}{2}$ XXZ spin chain with open boundary conditions

$$\begin{aligned}
\mathcal{H}_{XXZ} =& \sum_{j=1}^{L-1}\left[\sigma_j^x\sigma_{j+1}^x + \sigma_j^y\sigma_{j+1}^y + \operatorname{ch}\eta\,\sigma_j^z\sigma_{j+1}^z\right] + L\operatorname{ch}\eta \\
&+ \left[\sigma_1^z\coth\xi^- + \frac{2\kappa^-}{\operatorname{sh}\xi^-}\left(\sigma_1^x\operatorname{ch}\theta^- + i\sigma_1^y\operatorname{sh}\theta^-\right)\right]\operatorname{sh}\eta \\
&+ \left[\sigma_L^z\coth\xi^+ + \frac{2\kappa^+}{\operatorname{sh}\xi^+}\left(\sigma_L^x\operatorname{ch}\theta^+ + i\sigma_L^y\operatorname{sh}\theta^+\right)\right]\operatorname{sh}\eta
\end{aligned} \qquad (1.1)$$

and its limit to the spin-$\frac{1}{2}$ XXX chain via $\eta \to \epsilon ic$ for $\epsilon \to 0$

$$\begin{aligned}
\mathcal{H}_{XXX} =& \sum_{j=1}^{L-1}\left[\sigma_j^x\sigma_{j+1}^x + \sigma_j^y\sigma_{j+1}^y + \sigma_j^z\sigma_{j+1}^z\right] + L \\
&+ \left[\sigma_1^z\frac{ic}{\xi^-} + \frac{2ic\kappa^-}{\xi^-}\left(\sigma_1^x\operatorname{ch}\theta^- + i\sigma_1^y\operatorname{sh}\theta^-\right)\right] \\
&+ \left[\sigma_L^z\frac{ic}{\xi^+} + \frac{2ic\kappa^+}{\xi^+}\left(\sigma_L^x\operatorname{ch}\theta^+ + i\sigma_L^y\operatorname{sh}\theta^+\right)\right].
\end{aligned} \qquad (1.2)$$

The parameter η describes the anisotropy of the model. In the latter case the parameter c, introduced for the limit, only appears as a scale for the diagonal boundary field $\xi^\pm$. Note that for hermitian hamiltonians we need $\theta^\pm$ to be imaginary while the domain of $\xi^\pm$ depends on the value of the crossing parameter. For the XXX model and the massless case of the XXZ model, i.e. $\eta \in i\mathbb{R}$, we need $\xi^\pm$ to be purely imaginary. In the massive region of $\eta \in \mathbb{R}$ hermiticity demands $\xi^\pm$ to be real valued. These models, apart from being the simplest starting points for studies of boundary effects in a correlated system, allow to investigate the approach to a stationary state in one-dimensional diffusion problems for hard-core particles [22, 23] and transport through one-dimensional quantum systems [19].

In [83] Sklyanin himself solved the spectral problem for the XXZ model with *diagonal* boundary fields via the algebraic Bethe ansatz reproducing coordinate Bethe ansatz results by Alcaraz, Barber and Batchelor [3]. However the step to correlation functions without perturbational approaches was laid ground for only recently by Kitanine et al. [54] who solved the inverse problem, i.e. expressing local operators in terms of non-local elements of the algebra. In chapter 3 we will derive a non-linear integral equation for an auxiliary function describing the lowest lying state of zero magnetization and combine it with the known determinant formula of norms and scalar products [82]. In doing so, we extend the results of Kitanine from half infinite systems to arbitrary, including finite, system sizes.

1.2. Brief Historic Overview and Outline

The progress for the general case of (1.1) and (1.2) was hindered by the fact that the algebraic Bethe ansatz is not directly applicable due to the lack of a mandatory simple *pseudo vacuum* and alternative methods for the spectral problem must be developed. This task was approached in numerous ways.

First it is interesting to note that a suitable pseudo vacuum can be found by imposing certain constraints to be obeyed by the left and right boundary fields. With these constraints the eigenvalues of the spin-$\frac{1}{2}$ *XXZ* chain and of the isotropic spin-*S* model can be obtained by means of the algebraic Bethe ansatz [18, 68].

In a series of papers, Nepomechie *et al.* have worked out an approach that finally resulted in unconstrained boundary parameters but a restricted anisotropy of the *XXZ* model. They have been able to derive Bethe type equations whose roots parametrize the eigenvalues of the Hamiltonian for special values of the anisotropy $\eta = i\pi/(p+1)$ with p a positive integer and where the transfer matrix obeys functional equations of finite order [72]. The approach relies on the periodicity of the underlying trigonometric *R*-matrix of the model which is missing in the rational limit $\eta \to 0$ of the isotropic *XXX* chain. For generic values of the anisotropy the spectral problem has been formulated as a functional equation assuming that a certain limit of the transfer matrices exists [99].

No constraints at all are needed in the derivation of a different set of recursion relations for the diagonalization of (1.1) based on the representation theory of the q-Onsager algebra [7]. Their equations reduce the complexity of the system but do not resemble Bethe equations and the thermodynamic limit is still out of reach [6].

The putatively best method without any need to restrict the parameters was recently introduced by Galleas [35]. His functional approach uses the Yang-Baxter algebra in a completely different way than other methods. Taking certain matrix elements of the transfer matrix involving the ferromagnetic pseudo vacuum and the unknown eigenstates he derives equations vaguely reminiscent of nested Bethe equations as the eigenvalue is parametrized by two sets of dependent parameters. Unfortunately for this solution no numerical schemes, like iteration schemes for Bethe equations, have been worked out and numerical studies are limited to small system sizes. In addition no information on the eigenstates can be elaborated, therefore further studies of correlation functions are so far out of reach.

In this work we approach the problem by means of different methods which circumvent the difficulties of the algebraic Bethe ansatz in the absence of a reference state. In chapter 4 we apply Sklyanin's functional Bethe ansatz (or separation of variables method) [84] to the eigenvalue problem and in doing so we formulate it using a suitably chosen representation of the underlying Yang-Baxter algebra on a space of certain functions [34]. Unfortunately we need to restrict ourselves to the *XXX* hamiltonian (1.2) to leave the boundary parameters general. As the algebraic concept is independent from its representations, it is possible to treat models arising from different representations analogously. Hence we apply the results to the spin-boson model, i.e. a two site model coupling a spin and a single bosonic degree of freedom to each other.

In chapter 5 we compare the method of separation of variables to Nepomechie's method for roots of unity within the simpler setting of the twisted periodic XXZ model as both methods are applicable [78].

Finally using a different approach we derive integral equations for the rational case from fusion hierarchies for the eigenvalue of the transfer matrix describing the lowest lying state with vanishing magnetization of the chain.

Chapter 2

Algebraic Principles

In this chapter we will introduce the underlying theoretical concepts of this work. First we show how the hamiltonians arise within the Quantum Inverse Scattering Method and discuss the algebraic Bethe ansatz as one of the standard approaches to solve of the eigenvalue problem. Second we will introduce the concept of fusion and see that the transfer matrix eigenvalue is part of an infinite hierarchical set of non-linear equations.

2.1 Quantum Inverse Scattering Method

In order to put the *XXZ* model (1.1) and the *XXX* model (1.2) on a sound algebraic footing we construct a commuting family of observables containing the hamiltonian using the Quantum Inverse Scattering Method (QISM) [62].

Generally speaking the QISM provides a scheme for the construction of integrable quantum models and their integrals of motion starting from the Yang-Baxter algebra (YBA)

$$R_{12}(\lambda-\mu)T_1(\lambda)T_2(\mu) = T_2(\mu)T_1(\lambda)R_{12}(\lambda-\mu). \tag{2.1}$$

Here the $T_j(\lambda)$ are matrices on the linear auxiliary space V_j with entries being the generators of the quadratic algebra of operators on a quantum space. The 'structure' constants of the algebra are arranged in a quantum R-matrix, $R_{jk}(\lambda) \in \text{End}(V_j \otimes V_k)$. This R-matrix determines the class of the generated integrable quantum model but to do so R itself has to be a solution of the Yang-Baxter equation

$$R_{12}(\lambda_1-\lambda_2)R_{13}(\lambda_1-\lambda_3)R_{23}(\lambda_2-\lambda_3) = R_{23}(\lambda_2-\lambda_3)R_{13}(\lambda_1-\lambda_3)R_{12}(\lambda_1-\lambda_2). \tag{2.2}$$

The Yang-Baxter equation arose also as a relation for the two-body scattering matrix and is in this sense visualized in figure 2.1 [30]. Within this picture the integrability of the model

2. Algebraic Principles

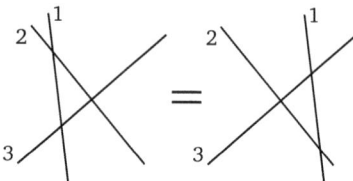

Figure 2.1: Graphical interpretation of the Yang-Baxter equation (2.2). Three particle scattering is broken up into two particle scattering – the order of processes being not important.

is assured through the fact that the full dynamics is described by non-diffractive two particle scattering. Note that the sequence of vertices in these kind of pictures is always determined from left to right, top to bottom.

For the models under consideration here we use the well-known trigonometric and rational solutions for two-dimensional spaces V_j corresponding to the 6-vertex model of classical statistical mechanics

$$R(\lambda,\mu) = \begin{pmatrix} a(\lambda,\mu) & 0 & 0 & 0 \\ 0 & b(\lambda,\mu) & c(\lambda,\mu) & 0 \\ 0 & c(\lambda,\mu) & b(\lambda,\mu) & 0 \\ 0 & 0 & 0 & a(\lambda,\mu) \end{pmatrix}, \qquad (2.3)$$

with

$$a(\lambda,\mu) = \mathrm{sh}(\lambda - \mu + \eta), \quad b(\lambda,\mu) = \mathrm{sh}(\lambda - \mu), \quad c(\lambda,\mu) = \mathrm{sh}\,\eta \qquad (2.4)$$

for the XXZ model and

$$a(\lambda,\mu) = \lambda - \mu + ic, \quad b(\lambda,\mu) = \lambda - \mu, \quad c(\lambda,\mu) = ic \qquad (2.5)$$

for the XXX model. Note that the parameter c does not have a physical meaning as it enters the hamiltonian (1.2) only as a scale for the boundary diagonal fields.

Besides the Yang-Baxter equation this R-matrix satisfies several other conditions such as symmetry with respect to the permutation operator P on $V \otimes V$ ($P x \otimes y = y \otimes x$),

$$R(\lambda) = PR(\lambda)P, \qquad (2.6)$$

unitarity involving some complex function $\rho(\lambda)$,

$$R(\lambda)R(-\lambda) = \rho(\lambda) \qquad (2.7)$$

and crossing unitarity for another complex function $\tilde{\rho}(\lambda)$,

$$R^{t_1}(\lambda)R^{t_1}(-\lambda - 2\eta) = \tilde{\rho}(\lambda). \qquad (2.8)$$

2.1. Quantum Inverse Scattering Method

Note that all functions for the XXX model can be obtained from those in the XXZ case in the rational limit, i.e. scaling the relevant parameters by a small ϵ and expanding around it

$$\lambda \to \epsilon\lambda, \quad \eta = \epsilon i c, \quad \epsilon \to 0. \tag{2.9}$$

This limit can be carried out at any point in the calculation of XXZ model to obtain the respective equations for the XXX model hence we will focus on the XXZ model to remain general when possible. Further note that (2.1) and (2.2) are independent of the normalization of the R-matrix. The complex functions in (2.7) and (2.8) are for the chosen normalization

$$\rho(\lambda) = \text{sh}(\lambda + \eta)\text{sh}(-\lambda + \eta) \quad, \quad \widetilde{\rho}(\lambda) = \text{sh}(-\lambda)\text{sh}(\lambda + 2\eta) \tag{2.10}$$

and respectively their rational limit.

Considering the trigonometric R-matrix (2.3) with (2.4) a representation of the Yang-Baxter algebra is the Lax operator

$$\mathcal{L}(\lambda) = \begin{pmatrix} \text{sh}(\lambda)S^0 + \text{ch}(\lambda)S^z & S^- \\ S^+ & \text{sh}(\lambda)S^0 - \text{ch}(\lambda)S^z \end{pmatrix}. \tag{2.11}$$

which describes a single quantum spin. For a spin-$\frac{1}{2}$ the elements of $\mathcal{L}$ are operators on a two-dimensional quantum space of states, a convenient representation via Pauli matrices is $S^0 = \text{ch}(\eta/2)\mathbb{1}$, $S^z = \text{sh}(\eta/2)\sigma_j^z$ and $S^\pm = \text{sh}(\eta/2)\text{ch}(\eta/2)\sigma^\pm$. The rational limit of (2.11) gives the representation for the rational $\mathcal{L}$-operator.

Another representation can be found by noticing that (2.1) turns into (2.2) for $T_j(\lambda) = R_{j3}(\lambda)$. For this reason the representation

$$\mathcal{L}(\lambda) = \phi(\lambda, \mu)R(\lambda, \mu). \tag{2.12}$$

of (2.1) is called 'fundamental' representation of the Yang-Baxter algebra. Clearly this representation can only be found if the auxiliary and quantum space are of same dimension. With a two-dimensional auxiliary space the fundamental representation describes spin-$\frac{1}{2}$ only. The normalization $\phi(\lambda, \mu)$ is chosen such that $\mathcal{L}$ turns into a permutation operator P for $\lambda = \mu$. This special value of the spectral parameter is called shift point and the existence of it guarantees the interactions in the hamiltonian to include only next-neighbouring spins.

There exists an algebra homomorphism for the Yang-Baxter algebra called co-multiplication by which products of representations of (2.1) are again representations of the algebra. Hence it is easily possible to describe multiple spins in different quantum spaces, e.g. different lattice sites or impurities. For L spins (2.11) yields

$$T(\lambda) = \mathcal{L}_L(\lambda)\mathcal{L}_{L-1}(\lambda)\cdots\mathcal{L}_1(\lambda) \tag{2.13}$$

where the index of the Lax operators refers to the quantum space the spin operators in (2.11) act on non-trivially. In the case of the fundamental representation embedding the R-matrix in the entire quantum space gives

$$T(\lambda) = R_{0L}(\lambda)R_{0(L-1)}(\lambda)\cdots R_{01}(\lambda) \tag{2.14}$$

2. Algebraic Principles

with the index 0 denoting the auxiliary space. To clarify the notation

$$R_{01}(\lambda) = R(\lambda) \otimes \mathbb{1}_2^{\otimes(L-1)} \tag{2.15}$$

$$R_{0j}(\lambda) = P_{1j} R_{01}(\lambda) P_{1j} \tag{2.16}$$

where P_{1j} is the permutation operator of the first and jth quantum space.

At this point we can already define a commuting family of operators which contains a physical interesting spin chain hamiltonian – however the boundary conditions are periodic. Upon tracing out the auxiliary space we arrive at an operator called transfer matrix acting on the combined quantum spaces

$$t_{\text{periodic}}(\lambda) = \text{tr}\, T(\lambda), \tag{2.17}$$

with $[t_{\text{periodic}}(\lambda), t_{\text{periodic}}(\mu)] = 0$. The hamiltonian is obtained as the logarithmic derivative of the transfer matrix at the shift point and its properties can be studied by using the entries of the monodromy matrix, i.e. the generators of the Yang-Baxter algebra

$$T(\lambda) = \begin{pmatrix} A(\lambda) & B(\lambda) \\ C(\lambda) & D(\lambda) \end{pmatrix}. \tag{2.18}$$

The commutations relations of the entries are given by (2.1) and are called fundamental commutation relations.

Another key element of the algebra is its central element or casimir [62]. It will not only be needed in the construction of open boundaries but it is a central ingredient for the methods based on the fusion hierarchy (see section 2.3). The composition of the casimir has a striking similarity to the 2×2 determinant of the monodromy matrix and hence is usually called quantum determinant $(d_q T)(u)$. With the one-dimensional projector P_{12}^- onto the antisymmetric (singlet) state in the tensor product $V \otimes V$ of auxiliary spaces the definition reads

$$\begin{aligned}(d_q T)(\lambda) &= \text{tr}_{12}\{P_{12}^- T_1(\lambda - \tfrac{\eta}{2}) T_2(\lambda + \tfrac{\eta}{2})\} \\ &= A(\lambda + \tfrac{\eta}{2}) D(\lambda - \tfrac{\eta}{2}) - B(\lambda + \tfrac{\eta}{2}) C(\lambda - \tfrac{\eta}{2}).\end{aligned} \tag{2.19}$$

Here, the trace tr_{12} is to be taken in both auxiliary spaces 1 and 2 of the tensor product $V \otimes V$. It is easy to show that the quantum determinant commutes with all generators as the R-matrix becomes a one-dimensional projector for $\mu = \lambda - \eta$ and hence the commutation relations from (2.1) are significantly simplified [62]. Thus the quantum determinant is a complex valued function times the identity operator in the quantum space. A different way to express the definition (2.19) is

$$T(\lambda + \tfrac{\eta}{2}) \sigma^y T^t(\lambda - \tfrac{\eta}{2}) \sigma^y = (d_q T)(\lambda) \cdot \mathbb{1} \tag{2.20}$$

and directly the inversion formula for the monodromy matrix

$$T^{-1}(\lambda) = \frac{1}{(d_q T)(\lambda - \tfrac{\eta}{2})} \sigma^y T^t(\lambda - \eta) \sigma^y, \tag{2.21}$$

and the factorization of the quantum determinant follow

$$(d_q T)(\lambda) = \prod_{j=1}^{L} (d_q \mathcal{L}_j)(\lambda) .\tag{2.22}$$

The latter allows to simply calculate its value from a single Lax operator. In the fundamental representation it is

$$(d_q T)(\lambda) = a(\lambda + \tfrac{\eta}{2})^L b(\lambda - \tfrac{\eta}{2})^L .\tag{2.23}$$

Example. *Consider the inhomogeneous periodic XXZ chain with inhomogeneities $s_j \in \mathbb{C}$ at each lattice site $j = 1\ldots L$. Then the quantum determinant of a fundamental Lax operator $\mathcal{L}_j(\lambda - s_j)$ (2.12) takes the scalar value*

$$(d_q \mathcal{L}_j)(\lambda) = \operatorname{sh}(\lambda - s_j + \tfrac{3\eta}{2}) \operatorname{sh}(\lambda - s_j - \tfrac{\eta}{2}) \tag{2.24}$$

yielding $(d_q T)(\lambda - \tfrac{\eta}{2}) = \left[\prod_{j=1}^{L} \operatorname{sh}(\lambda - s_j + \eta) \operatorname{sh}(\lambda - s_j - \eta) \right]$ for a chain of L local spins $\tfrac{1}{2}$.

2.1.1 Open boundary conditions within QISM

The extension to an algebraic background for open spin chains was developed by Sklyanin [83]. It is based on the representations of two algebras $\mathcal{T}^{(+)}$ and $\mathcal{T}^{(-)}$ defined by the relations

$$R_{12}(\lambda - \mu) \mathcal{T}_1^{(-)}(\lambda) R_{12}(\lambda + \mu) \mathcal{T}_2^{(-)}(\mu) = \mathcal{T}_2^{(-)}(\mu) R_{12}(\lambda + \mu) \mathcal{T}_1^{(-)}(\lambda) R_{12}(\lambda - \mu) \tag{2.25}$$

$$R_{12}(-\lambda + \mu) \mathcal{T}_1^{(+)t_1}(\lambda) R_{12}(-\lambda - \mu - 2\eta) \mathcal{T}_2^{(+)t_2}(\mu) = \\ = \mathcal{T}_2^{(+)t_2}(\mu) R_{12}(-\lambda - \mu - 2\eta) \mathcal{T}_1^{(+)t_1}(\lambda) R_{12}(-\lambda + \mu) .\tag{2.26}$$

The superscripts t_1 and t_2 denote transpositions with respect to the auxiliary spaces 1 and 2. We shall call $\mathcal{T}^{(+)}$ and $\mathcal{T}^{(-)}$ right and left reflection algebras respectively. The transfer matrix in this case is defined by

$$t(\lambda) = \operatorname{tr} \mathcal{T}^{(+)}(\lambda) \mathcal{T}^{(-)}(\lambda) \tag{2.27}$$

as a trace in auxiliary space. Again it generates with $[t(\lambda), t(\mu)] = 0$ a commuting family of operators and hence is the central object under consideration.

The explicit construction of integrable open boundary conditions for models arising from the Yang-Baxter algebra (2.1) starts with the 2×2 matrix

$$K(\lambda, \pm) = \frac{1}{\operatorname{sh}\xi^{\pm} \operatorname{ch}\lambda} \begin{pmatrix} \operatorname{sh}(\lambda + \xi^{\pm}) & \kappa^{\pm} e^{\theta^{\pm}} \operatorname{sh}(2\lambda) \\ \kappa^{\pm} e^{-\theta^{\pm}} \operatorname{sh}(2\lambda) & -\operatorname{sh}(\lambda - \xi^{\pm}) \end{pmatrix}$$
$$= \mathbb{1}_2 + \tanh\lambda \coth\xi^{\pm} \sigma^z + \frac{2\kappa^{\pm} e^{\theta^{\pm}} \operatorname{sh}\lambda}{\operatorname{sh}\xi^{\pm}} \sigma^+ + \frac{2\kappa^{\pm} e^{-\theta^{\pm}} \operatorname{sh}\lambda}{\operatorname{sh}\xi^{\pm}} \sigma^- \tag{2.28}$$

2. Algebraic Principles

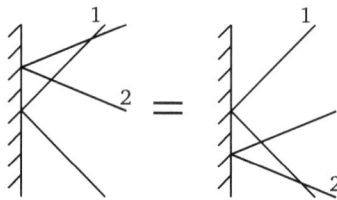

Figure 2.2: Visualization of the left reflection algebra (2.25) as inter-particle and wall scattering.

originally found by de Vega et al. [24]. A sometimes more suitable parametrization is obtained via [76]

$$\operatorname{sh}\alpha\operatorname{ch}\beta = \frac{\operatorname{sh}\xi}{2\kappa}, \quad \operatorname{ch}\alpha\operatorname{sh}\beta = \frac{\operatorname{ch}\xi}{2\kappa} \qquad (2.29)$$

and turns (2.28) into

$$K(\lambda,\pm) = \frac{1}{\operatorname{sh}\alpha^\pm \operatorname{ch}\beta^\pm \operatorname{ch}\lambda} \times$$
$$\begin{pmatrix} \operatorname{sh}\alpha^\pm \operatorname{ch}\beta^\pm \operatorname{ch}\lambda + \operatorname{ch}\alpha^\pm \operatorname{sh}\beta^\pm \operatorname{sh}\lambda & e^{\theta^\pm}\operatorname{sh}\lambda\operatorname{ch}\lambda \\ e^{-\theta^\pm}\operatorname{sh}\lambda\operatorname{ch}\lambda & \operatorname{sh}\alpha^\pm \operatorname{ch}\beta^\pm \operatorname{ch}\lambda - \operatorname{ch}\alpha^\pm \operatorname{sh}\beta^\pm \operatorname{sh}\lambda \end{pmatrix}. \qquad (2.30)$$

Each constitutes the known c-number representations $K^{(+)}(\lambda) = \frac{1}{2}K(\lambda+\eta,+)$ and $K^{(-)}(\lambda) = K(\lambda,-)$ of the reflection algebras with the obvious properties

$$\operatorname{tr} K(\lambda,\pm) = 2, \quad K^{(-)}(0) = \mathbb{1}, \quad \operatorname{tr} K^{(+)}(0) = 1. \qquad (2.31)$$

Note that applying the rational limit (2.9) to the boundary matrices the diagonal boundary parameters $\xi^\pm$ and $\alpha^\pm$ need to be rescaled as well whereas especially $\beta^\pm$ remains unscaled. This immediately follows from (2.29).

These special representations of the reflection algebras allow for a graphical interpretation which legitimates the name of the algebra. Viewing the K-matrix as a boundary and the R-matrices as two particles the order of reflection at the boundary and the inter-particle scattering is irrelevant. A visualization for the left reflection algebra is given in figure 2.2.

Operator valued representations of the reflection algebras can be constructed by again considering local $\mathscr{L}$-matrices building up two representations $T^{(+)}(\lambda) = \mathscr{L}_L(\lambda)\cdots\mathscr{L}_{M+1}(\lambda)$ and $T^{(-)}(\lambda) = \mathscr{L}_M(\lambda)\cdots\mathscr{L}_1(\lambda)$ of the Yang-Baxter algebra (2.1). Then by construction

$$\begin{aligned}\mathscr{T}^{(-)}(\lambda) &= T^{(-)}(\lambda)K^{(-)}(\lambda)T^{(-)-1}(-\lambda) \\ \mathscr{T}^{(+)t}(\lambda) &= T^{(+)t}(\lambda)K^{(+)}(\lambda)\bigl(T^{(+)-1}\bigr)^t(-\lambda)\end{aligned} \qquad (2.32)$$

are representations of the reflection algebras such that the normalized transfer matrix

$$t(\lambda) = \operatorname{tr} K^{(+)}(\lambda)T(\lambda)K^{(-)}(\lambda)T^{-1}(-\lambda), \quad t(0) = 1 \qquad (2.33)$$

2.1. Quantum Inverse Scattering Method

is independent of the factorization of $T(\lambda) = T^{(+)}(\lambda)T^{(-)}(\lambda)$. Thus we are free to choose to have the operator content in the left reflection algebra

$$\mathcal{T}^{(+)}(\lambda) = K^{(+)}(\lambda) \quad , \quad \mathcal{T}^{(-)}(\lambda) = T(\lambda)K^{(-)}(\lambda)T^{-1}(-\lambda) \,. \tag{2.34}$$

The inversion in the representation of the reflection algebra can be avoided by the inversion formula (2.21) including the quantum determinant of the Yang-Baxter algebra. In order to gain more symmetric arguments and to avoid inconvenient scalar functions after applying the inversion formula (2.21) it is instructive to define the new object $U(\lambda + \frac{\eta}{2}) \equiv \mathcal{T}^{(-)}(\lambda)(d_q T)(-\lambda - \frac{\eta}{2})$ consisting of

$$U(\lambda) = T(\lambda - \frac{\eta}{2})K^{(-)}(\lambda - \frac{\eta}{2})\sigma^y T^t(-\lambda - \frac{\eta}{2})\sigma^y \,. \tag{2.35}$$

It is still a representation of the left reflection algebra with a 2×2 matrix in auxiliary space

$$U(\lambda) = \begin{pmatrix} \mathcal{A}(\lambda) & \mathcal{B}(\lambda) \\ \mathcal{C}(\lambda) & \mathcal{D}(\lambda) \end{pmatrix} \,. \tag{2.36}$$

Using this representation we define a more suitable transfer matrix

$$\tau(\lambda) = (d_q T)(-\lambda)t(\lambda - \frac{\eta}{2}) = \operatorname{tr} K^+(\lambda + \frac{\eta}{2})U(\lambda) \,. \tag{2.37}$$

With the operators $\mathcal{A}, \mathcal{B}, \mathcal{C}$ and $\mathcal{D}$ the hamiltonian (1.1) can be studied as it is connected to the first derivative of $t(\lambda)$ by looking at the expansion $t(\lambda) = 1 + \lambda \frac{1}{\operatorname{sh}\eta} \mathcal{H}_{XXZ} + \ldots$ around the point $\lambda = 0$ or to the logarithmic derivative of $\tau(\lambda)$ at $\lambda = \frac{\eta}{2}$:

$$\begin{aligned} \mathcal{H}_{XXZ} &= \operatorname{sh}\eta\, t'(0) = \operatorname{sh}\eta \left(\partial_\lambda \ln \tau(\lambda) - \partial_\lambda \ln(d_q T)(-\lambda) \right)\Big|_{\lambda = \frac{\eta}{2}} \\ &= \operatorname{sh}\eta\, \partial_\lambda \ln \tau(\tfrac{\eta}{2}) \,. \end{aligned} \tag{2.38}$$

It is also possible to define quantum determinants for the reflection algebra as in the case of the Yang-Baxter algebras. We will concentrate on the left reflection algebra first. Its quantum determinant is defined by

$$(\Delta_q U)(\lambda) = \operatorname{tr}_{12} P_{12}^- U_1(\lambda - \tfrac{\eta}{2})R_{12}(2\lambda - \eta)U_2(\lambda + \tfrac{\eta}{2}) \,. \tag{2.39}$$

To express $(\Delta_q U)(\lambda)$ in terms of the generators $\mathcal{A}(\lambda), \mathcal{B}(\lambda), \mathcal{C}(\lambda)$ and $\mathcal{D}(\lambda)$ it is instructive to use the combinations

$$\widetilde{\mathcal{D}}(\lambda) \equiv \operatorname{sh}(2\lambda)\mathcal{D}(\lambda) - \operatorname{sh}\eta\, \mathcal{A}(\lambda), \quad \widetilde{\mathcal{C}}(\lambda) \equiv \operatorname{sh}(2\lambda + \eta)\mathcal{C}(\lambda) \tag{2.40}$$

borrowed from the algebraic Bethe ansatz [83]. Then the suggestive form of the quantum determinant reads

$$(\Delta_q U)(\lambda) = \mathcal{A}(\lambda + \tfrac{\eta}{2})\widetilde{\mathcal{D}}(\lambda - \tfrac{\eta}{2}) - \mathcal{B}(\lambda + \tfrac{\eta}{2})\widetilde{\mathcal{C}}(\lambda - \tfrac{\eta}{2}) \,. \tag{2.41}$$

2. Algebraic Principles

In case of the c-number representation $K(\lambda - \frac{\eta}{2}, -)$ for $U(\lambda)$ connected to the left reflection algebra the relation

$$(\Delta_q K)(\lambda - \tfrac{\eta}{2}, -) = \frac{\operatorname{sh}(2\lambda - 2\eta)\operatorname{ch}\lambda}{\operatorname{ch}(\lambda - \eta)} \det K(\lambda, -) \tag{2.42}$$

holds. Note that this relation is only valid for the shifted argument $\lambda - \frac{\eta}{2}$ because the arising expressions are no longer of difference form. For the boundary matrix chosen as in (2.30) the determinant $\det K(\lambda, -)$ factorizes and its quantum version decomposes to product form

$$(\Delta_q K)(\lambda - \tfrac{\eta}{2}, -) = -\operatorname{sh}(2\lambda - 2\eta)$$
$$\times \frac{\operatorname{sh}(\lambda - \alpha^-)\operatorname{ch}(\lambda - \beta^-)}{\operatorname{sh}\alpha^- \operatorname{ch}\beta^- \operatorname{ch}(\lambda - \eta)} \frac{\operatorname{sh}(\lambda + \alpha^-)\operatorname{ch}(\lambda + \beta^-)}{\operatorname{sh}\alpha^- \operatorname{ch}\beta^- \operatorname{ch}\lambda}. \tag{2.43}$$

It is obvious that in this parametrization (2.29) the model is invariant under the simultaneous transformations $\alpha \to -\alpha$ and $\beta \to i\pi - \beta$.

As the quantum determinant respects co-multiplication, applying it to the full representation (2.35) of the left reflection algebra with monodromy matrices T yields

$$(\Delta_q U)(\lambda) = (d_q T)(\lambda - \tfrac{\eta}{2})(\Delta_q K)(\lambda - \tfrac{\eta}{2}, -)(d_q T)(-\lambda - \tfrac{\eta}{2}). \tag{2.44}$$

As mentioned above the factorization of $T(\lambda) = T^+(\lambda)T^-(\lambda)$ is arbitrary and hence it is equivalently possible to shift all operator content to the right reflection algebra yielding instead of (2.34)

$$\mathcal{T}^{(+)t}(\lambda) = T^t(\lambda)K^{(+)t}(\lambda)(T^{-1})^t(\lambda) \quad , \quad \mathcal{T}^{(-)}(\lambda) = K^{(-)}(\lambda). \tag{2.45}$$

In this case the object for consideration is $U^{(+)t}(\lambda + \tfrac{\eta}{2}) \equiv \mathcal{T}^{(+)}(\lambda)(d_q T)(-\lambda - \tfrac{\eta}{2})$ explicitly given by

$$U^{(+)t}(\lambda) = T(\lambda - \tfrac{\eta}{2})K^{(-)}(\lambda - \tfrac{\eta}{2})\sigma^y T^t(-\lambda - \tfrac{\eta}{2})\sigma^y. \tag{2.46}$$

$U^{(+)}$ can also be written as a 2×2 matrix in auxiliary space

$$U^{(+)}(\lambda) = \begin{pmatrix} \mathcal{A}^{(+)}(\lambda) & \mathcal{B}^{(+)}(\lambda) \\ \mathcal{C}^{(+)}(\lambda) & \mathcal{D}^{(+)}(\lambda) \end{pmatrix}. \tag{2.47}$$

The transfer matrix $\tau(\lambda)$ under consideration is in this representation

$$\tau(\lambda) = \operatorname{tr} U^{(+)}(\lambda)K^-(\lambda - \tfrac{\eta}{2}) \tag{2.48}$$

and the Quantum Determinant of the right reflection algebra is defined by

$$(\Delta_q^+ U^{(+)})(\lambda) = \operatorname{tr}_{12} P_{12}^- U_2^{(+)t}(\lambda - \tfrac{\eta}{2})R_{12}(-2\lambda - \eta)U^{(+)t}{}_1(\lambda + \tfrac{\eta}{2}) \tag{2.49}$$
$$= \mathcal{D}^{(+)}(\lambda - \tfrac{\eta}{2}).\widetilde{\mathcal{A}}^{(+)}(\lambda - \tfrac{\eta}{2}) + \mathcal{B}^{(+)}(\lambda + \tfrac{\eta}{2})\widetilde{\mathcal{C}}^{(+)}(\lambda - \tfrac{\eta}{2}) \tag{2.50}$$

with

$$\widetilde{\mathcal{A}}^{(+)}(\lambda) \equiv \operatorname{sh}(-2\lambda).\mathcal{A}^{(+)}(\lambda) - \operatorname{sh}\eta\, \mathcal{D}^{(+)}(\lambda), \quad \widetilde{\mathcal{C}}^{(+)}(\lambda) \equiv \operatorname{sh}(-2\lambda + \eta)\mathcal{C}^{(+)}(\lambda). \tag{2.51}$$

The case of a c-number representation simplifies to

$$(\Delta_q^+ K)(\lambda + \tfrac{\eta}{2}, +) = -2\operatorname{sh}(\lambda + \eta)\operatorname{ch}\lambda \det K(\lambda, +). \tag{2.52}$$

2.2 Algebraic Bethe Ansatz

The standard choice to study a model generated by QISM is the algebraic Bethe ansatz. This method starts with a particular reference state or pseudo vacuum $|0\rangle$ obeying with scalar functions $a(\lambda)$ and $d(\lambda)$

$$C(\lambda)|0\rangle = 0, \quad A(\lambda)|0\rangle = a(\lambda)|0\rangle, \quad D(\lambda)|0\rangle = d(\lambda)|0\rangle, \tag{2.53}$$

here as an example for the monodromy matrix (2.18) of the Yang-Baxter algebra. Especially this means that $|0\rangle$ is an eigenstate of the corresponding transfer matrix and hence of the hamiltonian. In the case of the periodic XXZ model a simple reference state is the ferromagnetic state with all spins aligned with respect to the z-direction. Further eigenstates are then found by application of the other off-diagonal operator on the pseudo vacuum

$$|\psi(\{\lambda_j\})\rangle = \prod_{j=1}^{M} B(\lambda_j)|0\rangle. \tag{2.54}$$

The integer M is found to count the number of turned spins with respect to the reference state. The ordering of operators B is irrelevant as $[B(\lambda_i), B(\lambda_j)] = 0$ from (2.1) but the choice of the spectral parameters is restricted. In order for (2.54) to be an eigenstate, terms containing a $B(\mu)$ after commuting A and D to the right in

$$t(\mu)|\psi(\{\lambda_j\})\rangle = A(\mu)B(\lambda_1)\ldots B(\lambda_M)|0\rangle + D(\mu)B(\lambda_1)\ldots B(\lambda_M)|0\rangle \tag{2.55}$$

have to vanish. These conditions are called Bethe ansatz equations

$$\left[\frac{\operatorname{sh}(\lambda_j - \frac{\eta}{2})}{\operatorname{sh}(\lambda_j + \frac{\eta}{2})}\right]^{2L} = \left[\prod_{\substack{\ell=1 \\ \ell \neq j}}^{M} \frac{\operatorname{sh}(\lambda_j - \lambda_\ell - \eta)}{\operatorname{sh}(\lambda_j - \lambda_\ell + \eta)}\right]. \tag{2.56}$$

The corresponding eigenvalue of the transfer matrix t_{periodic} (2.17) can then be read off (2.55)

$$\Lambda_{\text{periodic}}(\lambda) = \operatorname{sh}^L(\lambda + \tfrac{\eta}{2}) \left[\prod_{\ell=1}^{M} \frac{\operatorname{sh}(\lambda - \lambda_\ell - \eta)}{\operatorname{sh}(\lambda - \lambda_\ell)}\right] + \operatorname{sh}^L(\lambda - \tfrac{\eta}{2}) \left[\prod_{\ell=1}^{M} \frac{\operatorname{sh}(\lambda - \lambda_\ell + \eta)}{\operatorname{sh}(\lambda - \lambda_\ell)}\right]. \tag{2.57}$$

In the case of open boundary conditions this procedure is also applicable for diagonal boundary fields, i.e. $\kappa^\pm = 0$ in (2.28) or $\beta^\pm \to \infty$ in (2.30) respectively. The reference state is again a ferromagnetically aligned state but it is convenient to use $\widetilde{\mathscr{D}}(\lambda)$ and $\mathscr{A}^{(+)}(\lambda)$ from (2.40) and (2.51) to express the transfer matrix $\tau(\lambda)$ from (2.37) and (2.48) respectively as then the commutation relations with $\mathscr{B}$ or $\mathscr{B}^{(+)}$ significantly simplify [83]. An eigenstate is characterized by Bethe numbers $\{\lambda_j\}_{j=1}^{M}$ fulfilling the Bethe ansatz equations

$$\frac{\operatorname{sh}(\lambda_j - \xi^+ + \tfrac{\eta}{2})\operatorname{sh}(\lambda_j - \xi^- + \tfrac{\eta}{2})}{\operatorname{sh}(\lambda_j + \xi^+ - \tfrac{\eta}{2})\operatorname{sh}(\lambda_j + \xi^- - \tfrac{\eta}{2})} \left[\frac{\operatorname{sh}(\lambda_j - \tfrac{\eta}{2})}{\operatorname{sh}(\lambda_j + \tfrac{\eta}{2})}\right]^{2L} = \left[\prod_{\substack{\ell=1 \\ \ell \neq j}}^{M} \frac{\operatorname{sh}(\lambda_j - \lambda_\ell - \eta)\operatorname{sh}(\lambda_j + \lambda_\ell - \eta)}{\operatorname{sh}(\lambda_j - \lambda_\ell + \eta)\operatorname{sh}(\lambda_j + \lambda_\ell + \eta)}\right]. \tag{2.58}$$

2. Algebraic Principles

Within this framework the state itself is then constructed using off-diagonal operators of either choices of representations as the generated states are proportional to each other [54]

$$|\psi(\{\lambda_j\})\rangle = \prod_{j=1}^{M} \mathcal{B}(\lambda_j)|0\rangle \sim \prod_{j=1}^{M} \mathcal{B}^{(+)}(\lambda_j)|0\rangle . \tag{2.59}$$

The eigenvalue to the state (2.59) is then given by

$$\Lambda(\lambda) = \frac{(-1)^L \operatorname{sh}^{2L}(\lambda + \frac{\eta}{2})}{2 \operatorname{ch}(\lambda + \frac{\eta}{2}) \operatorname{ch}(\lambda - \frac{\eta}{2})} \frac{\operatorname{sh}(2\lambda + \eta)}{\operatorname{sh}(2\lambda)} \frac{\operatorname{sh}(\lambda + \alpha^+ - \frac{\eta}{2})}{\operatorname{sh}\alpha^+} \frac{\operatorname{sh}(\lambda + \alpha^- - \frac{\eta}{2})}{\operatorname{sh}\alpha^-} \frac{q(\lambda - \eta)}{q(\lambda)}$$
$$+ \frac{(-1)^L \operatorname{sh}^{2L}(\lambda - \frac{\eta}{2})}{2 \operatorname{ch}(\lambda + \frac{\eta}{2}) \operatorname{ch}(\lambda - \frac{\eta}{2})} \frac{\operatorname{sh}(2\lambda - \eta)}{\operatorname{sh}(2\lambda)} \frac{\operatorname{sh}(\lambda - \alpha^+ + \frac{\eta}{2})}{\operatorname{sh}\alpha^+} \frac{\operatorname{sh}(\lambda - \alpha^- + \frac{\eta}{2})}{\operatorname{sh}\alpha^-} \frac{q(\lambda + \eta)}{q(\lambda)} \tag{2.60}$$

where $q(\lambda) \equiv \left[\prod_{\ell=1}^{M} \operatorname{sh}(\lambda - \lambda_\ell) \operatorname{sh}(\lambda - \lambda_\ell)\right]$.

For open boundary conditions there is no easy pseudo vacuum known as the number of turned spins is not a good quantum number of the hamiltonian (1.1) and hence the completely ferromagnetic state is not an eigenstate of the transfer matrix. There have been attempts to construct different pseudo vacua (e.g. [18, 68]) but the solutions obtained are only valid for constricted boundary parameters.

Some other approaches to the eigenvalue problem are based on functional relations of the eigenvalue as a function of the spectral parameter. These include the functional Bethe ansatz and all approaches using so-called fusion hierarchies. Hence we will discuss the concept of fusion in the next section.

2.3 Fusion in Auxiliary Space

The so-called *fusion* procedure grants the possibility to easily obtain R-matrices and boundary matrices of higher dimension obeying a Yang-Baxter equation or a reflection equation respectively. It is called fusion as e.g. two objects with spin-$\frac{1}{2}$ auxiliary spaces are combined to one object with spin-1 auxiliary space. In case of the R-matrix this procedure is applicable to the auxiliary, the quantum space, and even both. Furthermore the associated transfer matrices are not independent from each other but satisfy functional relations called *fusion hierarchies*. From these hierarchies the ultimate goal is to obtain the desired eigenvalue of the respective spin chains.

Besides an explicit derivation of the first levels of the following fusion equations one can apply the powerful graphical scheme to interpret, read and proof the equations. Basics parts of this scheme have been introduced in figures 2.1 and 2.2 for the Yang-Baxter equation and reflection equation respectively. The graphical method often provides motivation and hints regarding calculations in almost all aspects of integrable models (see e.g. [47, 58]).

Since this work concerns quantum spins-$\frac{1}{2}$ the fusion in auxiliary spaces is focussed. The fusion of R-matrices was developed by Kulish, Reshetikhin and Sklyanin [64, 65]. The

2.3. Fusion in Auxiliary Space

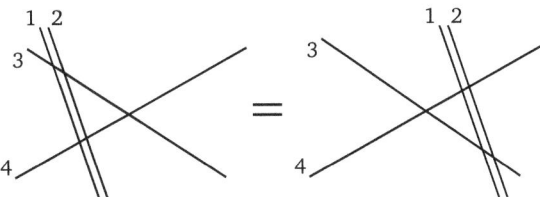

Figure 2.3: Visualization of (2.61). Two YBE serve as a basis to fuse spaces 1 and 2.

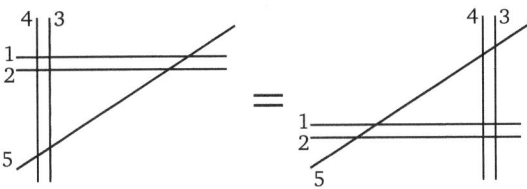

Figure 2.4: The YBA involves two auxiliary spaces. This figure represents it for fused auxiliary spaces 1 and 2, and 3 and 4.

procedure for fusion in one space starts with the equation

$$(R_{13}R_{23})(R_{14}R_{24})R_{34} = R_{34}(R_{14}R_{24})(R_{13}R_{23}) \tag{2.61}$$

depicted in figure 2.3. We used the abbreviation $R_{ij} \equiv R_{ij}(\lambda_i - \lambda_j)$ and its validity is clear from Yang-Baxter equations for the appropriate spaces and spectral parameters. The figure already suggests that spaces 1 and 2 will be fused to give a spin-1 auxiliary space. Another key element is the so-called *triangularity condition*

$$P_{12}^- R_{13}(\tau - \eta) R_{23}(\tau) P_{12}^+ = 0 \,. \tag{2.62}$$

It arises from the Yang-Baxter equation (2.2) by setting $\lambda_1 - \lambda_2 = -\eta$ so that $R_{12}(-\eta) = P_{12}^-$ is the projector on the singlet and multiplication from the right with the projector on the triplet P_{12}^+. Arranging the arguments in equation (2.61) properly to use the triangularity condition and defining the fused R-matrix as

$$R_{(12)3}(\lambda - \eta) = P_{12}^+ R_{13}(\lambda - \eta) R_{23}(\lambda) P_{12}^+ \tag{2.63}$$

equation (2.61) becomes a Yang-Baxter equation for the fused R-matrix

$$R_{(12)3}(\lambda) R_{(12)4}(\lambda + \mu) R_{34}(\mu) = R_{34}(\mu) R_{(12)4}(\lambda + \mu) R_{(12)3}(\lambda) \,. \tag{2.64}$$

To see that this R-matrix is a representation of a Yang-Baxter algebra we consider figure 2.4 which represents the equation

$$(R_{14}R_{13}R_{24}R_{23})(R_{15}R_{25})(R_{35}R_{45}) = (R_{35}R_{45})(R_{15}R_{25})(R_{14}R_{13}R_{24}R_{23}) \tag{2.65}$$

2. Algebraic Principles

again with the abbreviation $R_{ij} = R_{ij}(\lambda_i - \lambda_j)$. Fixing the arguments to use the triangularity (2.62) again and introducing the definition of the R-matrix fused in both spaces

$$R_{\langle 12\rangle\langle 34\rangle}(\lambda) = P_{12}^+ P_{34}^+ R_{14}(\lambda - \eta) R_{24}(\lambda) R_{23}(\lambda + \eta) P_{12}^+ P_{34}^+ \tag{2.66}$$

we find the desired Yang-Baxter equation that reveals (2.63) as a representation of a Yang-Baxter algebra with 'structure' constants $R_{\langle 12\rangle\langle 34\rangle}$

$$R_{\langle 12\rangle\langle 34\rangle}(\lambda - \mu) R_{\langle 12\rangle 5}(\lambda) R_{\langle 34\rangle 5}(\mu) = R_{\langle 34\rangle 5}(\mu) R_{\langle 12\rangle 5}(\lambda) R_{\langle 12\rangle\langle 34\rangle}(\lambda - \mu) . \tag{2.67}$$

The associated transfer matrix analogously constructed to the periodic chain is then $t_2^{\text{YBA}}(\lambda) \equiv \text{tr}_{\langle 12\rangle} R_{\langle 12\rangle L} R_{\langle 12\rangle L-1} \ldots R_{\langle 12\rangle 1}$. As mentioned in the beginning the fusion process is also applicable to the second space of the R-matrix resulting in

$$R_{1\langle 23\rangle}(\lambda) = P_{23}^+ R_{13}(\lambda - \eta) R_{23}(\lambda) P_{23}^+ \tag{2.68}$$

from which one can not only built spin chains with higher quantum spin [51, 53] but it will also be needed for the fusion of representations of the reflection algebras.

The scheme showed here for two two-dimensional spaces can be carried further fusing up to an arbitrary high dimensional auxiliary space. The emerging representation for a spin-k auxiliary space and spin-$\frac{1}{2}$ quantum space denoted by ℓ is [64, 65, 75]

$$R_{\langle 1\ldots k\rangle\ell}(\lambda) = P_{1\ldots k}^+ R_{1\ell}(\lambda) R_{2,\ell}(\lambda + \eta) \cdots R_{k,\ell}(\lambda + (k-1)\eta) P_{1\ldots k}^+ . \tag{2.69}$$

The appearing projector P^+ is defined by

$$P_{1\ldots n}^+ = \frac{1}{n!} \sum_\sigma P_\sigma \tag{2.70}$$

where the sum is over all permutations $\sigma = (\sigma_1, \sigma_2, \ldots, \sigma_n)$ of $(1, 2, \ldots, n)$ and P_σ is the permutation operator in the space $\mathbb{C}_2^{\otimes n}$. For instance,

$$\begin{aligned} P_{12}^+ &= \tfrac{1}{2}(\mathbb{1} + P_{12}) \\ P_{123}^+ &= \tfrac{1}{6}(\mathbb{1} + P_{23} P_{12} + P_{12} P_{23} + P_{12} + P_{23} + P_{13}) . \end{aligned} \tag{2.71}$$

The associated transfer matrix is a trace over the complete auxiliary space

$$t_k^{\text{YBA}} \equiv \text{tr}_{\langle 1\ldots k\rangle} R_{\langle 1\ldots k\rangle L} \ldots R_{\langle 1\ldots k\rangle 1} . \tag{2.72}$$

For these transfer matrices the following fusion hierarchy with $k = 1, 2, 3, \ldots$ holds [52, 53, 65, 75]

$$t_{k+1}^{\text{YBA}}(\lambda) = t_k^{\text{YBA}}(\lambda) t_1^{\text{YBA}}(\lambda + k\eta) - (d_q T)(\lambda + (k-1)\eta) t_{k-1}^{\text{YBA}}(\lambda) \quad , \tag{2.73}$$

where $t_1^{\text{YBA}}(\lambda) \equiv t_{\text{periodic}}(\lambda)$ and $t_0^{\text{YBA}}(\lambda) \equiv \mathbb{1}$.

In the open boundary case the procedure needs to be extended to the reflection algebras [69]. Again we will start with scattering processes as graphical interpretations of equations.

2.3. Fusion in Auxiliary Space

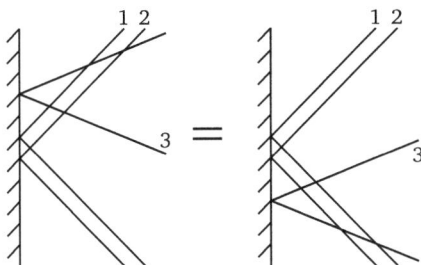

Figure 2.5: Graphical interpretation of the starting equation (2.74) for fusion of the left reflection algebra.

Adding $K_i \equiv K_i(\lambda_i)$ to the abbreviations we have for the left reflection algebra from figure 2.5

$$R_{32}R_{31}K_3R_{31}R_{32}K_1R_{12}K_2 = K_1R_{21}K_2R_{31}R_{32}K_3R_{32}R_{31} \,. \tag{2.74}$$

By setting $\lambda - \mu = -\eta$ in (2.25) we find a triangularity condition for the left reflection algebra

$$P_{12}^- K_1(\lambda) R_{12}(2\lambda + \eta) K_2(\lambda + \eta) P_{12}^+ = 0 \,. \tag{2.75}$$

Defining the fused boundary matrix as

$$K_{\langle 12 \rangle}(\lambda) = P_{12}^+ K_1(\lambda) R_{12}(2\lambda + \eta) K_2(\lambda + \eta) P_{12}^+ \tag{2.76}$$

and using triangularity condition for the YBA (2.62) and the definition of the fused R-matrix (2.63) we find a left reflection equation for the fused boundary matrices

$$R_{3\langle 12 \rangle}(\mu - \lambda) K_3(\mu) R_{\langle 12 \rangle 3}(\mu + \lambda) K_{\langle 12 \rangle}(\lambda) = K_{12}(\lambda) R_{\langle 12 \rangle 3}(\mu + \lambda) K_3(\mu) R_{3\langle 12 \rangle}(\mu - \lambda) \,. \tag{2.77}$$

Analogously for the right algebra we have (an index at transposition denotes the space where it is performed)

$$\begin{aligned}R_{\langle 12 \rangle 3}^{t_{123}}(-\mu + \lambda) K_3^t(\mu) R_{3\langle 12 \rangle}^{t_{123}}(-\mu - \lambda - 2\eta) K_{\langle 12 \rangle}^t(\lambda) \\ = K_{\langle 12 \rangle}^t(\lambda) R_{3\langle 12 \rangle}^{t_{123}}(-\mu - \lambda - 2\eta) K_3^t(\mu) R_{\langle 12 \rangle 3}^{t_{123}}(-\mu + \lambda) \,.\end{aligned} \tag{2.78}$$

by using the definition

$$K_{\langle 12 \rangle}^t(\lambda) = P_{12}^+ K_1^t(\lambda) R_{12}(-2\lambda - 3\eta) K_2^t(\lambda + \eta) P_{12}^+ \,. \tag{2.79}$$

To construct the transfer matrix with a spin-1 auxiliary space we need an operator valued representation of at least one of the reflection algebras. As in (2.34) we choose to put the operator content in the left algebra. Utilizing the co-multiplication property we have

$$\mathcal{T}_{\langle 12 \rangle}^{(-)} = P_{12}^+ \mathcal{T}_1^{(-)} R_{12}(2\lambda + \eta) \mathcal{T}_2^{(-)}(\lambda + \eta) P_{12}^+ \tag{2.80}$$

2. Algebraic Principles

yielding the transfer matrix with a spin-1 auxiliary space $t_{(12)} \equiv \text{tr}_{(12)} K^+_{(12)} \mathcal{T}^{(-)}_{(12)}$. With this definitions Mezincescu and Nepomechie [69] showed the fusion formula for the transfer matrix of the open boundary model

$$t_{(12)}(\lambda - \eta) = -\text{sh}(2\lambda - \eta)\text{sh}(2\lambda + \eta)t(\lambda - \eta)t(\lambda) - \tfrac{1}{4}(\Delta_q^+ K)(\lambda + \tfrac{\eta}{2})(\Delta_q \mathcal{T}^{(-)})(\lambda) \quad (2.81)$$

depending on the quantum determinants of both reflection algebras and the original transfer matrix from (2.27). Changing this result to the representation U of the left reflection algebra and getting rid of the scalar factor by definition of $\tau_{(12)}$ we find

$$\tau_{(12)}(\lambda) = \tau(\lambda - \tfrac{\eta}{2})\tau(\lambda + \tfrac{\eta}{2}) - \delta(\lambda) \quad (2.82)$$

with the scalar function δ on the right hand side

$$\delta(\lambda) = \frac{\text{sh}(\lambda + \eta)\text{sh}(\lambda - \eta)}{\text{sh}(\lambda - 2\eta)\text{sh}(2\lambda + \eta)\text{ch}^2\lambda}(d_q T)(\lambda - \tfrac{\eta}{2})(d_q T)(-\lambda - \tfrac{\eta}{2})$$
$$\times \frac{\text{sh}(\lambda - \alpha^+)\text{ch}(\lambda - \beta^+)}{\text{sh}\alpha^+\text{ch}\beta^+}\frac{\text{sh}(\lambda + \alpha^+)\text{ch}(\lambda + \beta^+)}{\text{sh}\alpha^+\text{ch}\beta^+} \quad (2.83)$$
$$\times \frac{\text{sh}(\lambda - \alpha^-)\text{ch}(\lambda - \beta^-)}{\text{sh}\alpha^-\text{ch}\beta^-}\frac{\text{sh}(\lambda + \alpha^-)\text{ch}(\lambda + \beta^-)}{\text{sh}\alpha^-\text{ch}\beta^-}.$$

Defining $t_1(\lambda) \equiv \tau(\lambda + \tfrac{\eta}{2})$ and $t_2(\lambda) \equiv \tau_{(12)}(\lambda - \eta)$ leaves us with the more favorable form of

$$t_2(\lambda - \eta) = t_1(\lambda - \eta)t_1(\lambda) - \delta(\lambda) \quad (2.84)$$

for extending this procedure to higher dimensional auxiliary spaces [76, 102]. The arising transfer matrices t_k for integer k are finally related to each other through the fusion hierarchy

$$t_k(\lambda - (k-1)\eta) = t_{k-1}(\lambda - (k-1)\eta)t_1(\lambda) - \delta(\lambda)t_{k-2}(\lambda - (k-1)\eta). \quad (2.85)$$

2.3.1 Asymptotic of transfer matrices

The asymptotic behaviour of the eigenvalue of the spin-$\tfrac{1}{2}$ and the fused transfer matrices is easy to obtain in the rational limit but it is an important information for the approaches presented.

$$\left.\begin{array}{c}\tau(\lambda)\\t_1(\lambda)\end{array}\right\}\xrightarrow[\lambda\to\pm\infty]{}\frac{\text{ch}(\phi)}{\alpha^+\alpha^-}\lambda^{2L+2}\equiv\frac{(-1)^L\,\text{sh}\beta^+\text{sh}\beta^-+\text{ch}(\Theta^+-\Theta^-)}{\alpha^+\alpha^-\,\text{ch}\beta^+\text{ch}\beta^-}\lambda^{2L+2} \quad (2.86)$$

Observing the proportionality $t_k(\lambda) \sim a_k \lambda^{k(2L+2)}$ we can calculate the coefficients by solving the recursion relation

$$a_k = \frac{\text{ch}\phi}{\alpha^+\alpha^-}a_{k-1} - \frac{1}{(2\alpha^+\alpha^-)^2}a_{k-2}, \quad a_0 = 1, \quad a_1 = \frac{\text{ch}\phi}{\alpha^+\alpha^-} \quad (2.87)$$

and obtain

$$a_k = \frac{1}{(2\alpha^+\alpha^-)^k}\frac{\text{sh}((k+1)\phi)}{\text{sh}\phi}. \quad (2.88)$$

Chapter 3

Scalar Products for the Diagonal *XXZ* Spin Chain

In this chapter we study the *XXZ* model (1.1) with diagonal boundaries, i.e. $\kappa^\pm = 0$ or $\beta^\pm \to \infty$ respectively with the goal to calculate expectation values for finite chain lengths.

The model is well studied and can in addition to the algebraic Bethe ansatz (see section 2.2) even be solved by means of the coordinate Bethe ansatz [1]. The distribution of Bethe roots in the ground state depends on the boundary fields. Anisotropies $|\Delta| < 1$ for example allow for at most two purely imaginary Bethe numbers besides real roots [85]. From the coordinate Bethe ansatz such states containing imaginary rapidities are usually termed *boundary bound states*. This terminology is related to the exponential decay of phase factors. The ground state for anisotropies $\Delta > 1$ can be found in [49].

To calculate the e.g. S^z-magnetization in the ground state as an expectation value in the framework of the algebraic Bethe ansatz (see section 2.2) one could make use of the inverse problem Wang [95] solved in terms of a mixture of the reflection and the Yang-Baxter algebra. However, this method makes use of the translation operator of the periodic spin chain for which the Bethe states are no longer eigenstates. Kitanine *et al.* overcame this difficulty by reducing the problem to the algebra of the periodic chain only where the inverse problem [56] is only expressed in terms of its algebra. So they determined the action of local operators on Bethe vectors in the representation of the reflection algebra and thus were able to apply the trigonometric generalization [54] of the rational determinant formula [96] for scalar products. Additionally they succeeded in simplifying the combinatorial part of the local S^z-magnetization by introducing a generating function [55]. The result was a multiple integral representation for its state average value. The integral representation is linked to expressions from the vertex operator approach [46] and was derived for the ground state of (1.1) described by Bethe root densities which are valid in the thermodynamic limit $L \to \infty$ of the half infinite chain.

In order to study finite chains we will first need to introduce a non-linear integral equation (NLIE) for an auxiliary function accounting for the lowest lying state of the hamiltonian (1.1) with zero magnetization. Then we will extensively built on the above mentioned results of Kitanine et al. [54] and combine their determinant formula with the auxiliary function to represent normalized scalar products in terms of multiple integrals. This result can then be applied as an example to the generating function of the S^z-magnetization to obtain the expectation value for finite chain lengths.

3.1 Auxiliary Function

For the calculations in this chapter we will consider the inhomogeneous model as in Example containing eq. (2.24), thus comparisons have to be made in the homogeneous limit $s_j \to 0$. The Bethe Ansatz equations (2.58) for $M \leq L/2$ Bethe numbers and the corresponding eigenvalue (2.60) of the transfer matrix (2.37) are modified as well resulting in

$$\frac{\operatorname{sh}(\lambda_j - \xi^+ + \frac{\eta}{2})\operatorname{sh}(\lambda_j - \xi^- + \frac{\eta}{2})}{\operatorname{sh}(\lambda_j + \xi^+ - \frac{\eta}{2})\operatorname{sh}(\lambda_j + \xi^- - \frac{\eta}{2})} \left[\prod_{\ell=1}^{L} \frac{\operatorname{sh}(\lambda_j - \frac{\eta}{2} + s_\ell)\operatorname{sh}(\lambda_j - \frac{\eta}{2} - s_\ell)}{\operatorname{sh}(\lambda_j + \frac{\eta}{2} + s_\ell)\operatorname{sh}(\lambda_j + \frac{\eta}{2} - s_\ell)} \right]$$
$$= \left[\prod_{\substack{\ell=1 \\ \ell \neq j}}^{M} \frac{\operatorname{sh}(\lambda_j - \lambda_\ell - \eta)\operatorname{sh}(\lambda_j + \lambda_\ell - \eta)}{\operatorname{sh}(\lambda_j - \lambda_\ell + \eta)\operatorname{sh}(\lambda_j + \lambda_\ell + \eta)} \right], \tag{3.1}$$

$$\Lambda(z) = \frac{(-1)^L \phi(z + \frac{\eta}{2})}{2\operatorname{ch}(z + \frac{\eta}{2})\operatorname{ch}(z - \frac{\eta}{2})}$$
$$\times \frac{\operatorname{sh}(2z + \eta)}{\operatorname{sh}(2z)} \frac{\operatorname{sh}(z + \xi^+ - \frac{\eta}{2})}{\operatorname{sh}\xi^+} \frac{\operatorname{sh}(z + \xi^- - \frac{\eta}{2})}{\operatorname{sh}\xi^-} \frac{q(z - \eta)}{q(z)}$$
$$+ \frac{(-1)^L \phi(z - \frac{\eta}{2})}{2\operatorname{ch}(z + \frac{\eta}{2})\operatorname{ch}(z - \frac{\eta}{2})} \tag{3.2}$$
$$\times \frac{\operatorname{sh}(2z - \eta)}{\operatorname{sh}(2z)} \frac{\operatorname{sh}(z - \xi^+ + \frac{\eta}{2})}{\operatorname{sh}\xi^+} \frac{\operatorname{sh}(z - \xi^- + \frac{\eta}{2})}{\operatorname{sh}\xi^-} \frac{q(z + \eta)}{q(z)}.$$

Although remembering $q(z) = \left[\prod_{\ell=1}^{M} \operatorname{sh}(z - \lambda_\ell)\operatorname{sh}(z + \lambda_\ell) \right]$ the eigenvalue is analytic at the Bethe roots λ_j due to Bethe Ansatz equation (3.1). The shorthand $\phi(z) \equiv \left[\prod_{\ell=1}^{L} \operatorname{sh}(z - s_\ell)\operatorname{sh}(z + s_\ell) \right]$ accounts for the pairwise distinct lattice inhomogeneities s_j regularizing combinatorial expressions in the forthcoming sections.

Let us restrict the anisotropy $\operatorname{ch}\eta$ of the zz-interaction to the massless case, $\eta = i\gamma$, and choose $0 < \gamma < \pi/2$ for the region next to the isotropic point. By selecting the lowest lying state of zero magnetization, not necessarily the ground state, from Bethe vectors

3.1. Auxiliary Function

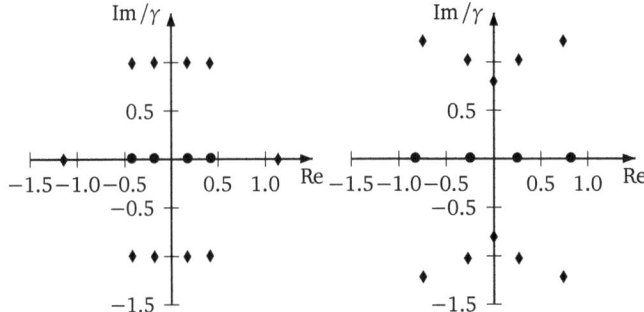

Figure 3.1: Solution of $1+\mathfrak{a}(z)=0$ in the rational limit for the ground state of $L=4$ lattice sites: A typical distribution of (two) Bethe roots ● and (five) hole-type solutions ♦ for positive boundary fields $\xi^+ = 1.1\mathrm{i}$, $\xi^- = .9\mathrm{i}$ (left panel) and a negative boundary field $\xi^- = -.3\mathrm{i}$ along with $\xi^+ = 4\mathrm{i}$ (right panel) in the sector of $M=2$.

$\mathcal{B}(\lambda_1)\ldots\mathcal{B}(\lambda_M)|0\rangle$ with $M=L/2$ some simplifications occur for the auxiliary function defined by

$$\mathfrak{a}(z) \equiv \frac{\operatorname{sh}(z-\xi^+ + \tfrac{\mathrm{i}\gamma}{2})\operatorname{sh}(z-\xi^- + \tfrac{\mathrm{i}\gamma}{2})\operatorname{sh}(2z-\mathrm{i}\gamma)}{\operatorname{sh}(z+\xi^+ - \tfrac{\mathrm{i}\gamma}{2})\operatorname{sh}(z+\xi^- - \tfrac{\mathrm{i}\gamma}{2})\operatorname{sh}(2z+\mathrm{i}\gamma)} \\ \times \left[\prod_{\ell=1}^{L} \frac{\operatorname{sh}(z-\tfrac{\mathrm{i}\gamma}{2}+s_\ell)\operatorname{sh}(z-\tfrac{\mathrm{i}\gamma}{2}-s_\ell)}{\operatorname{sh}(z+\tfrac{\mathrm{i}\gamma}{2}+s_\ell)\operatorname{sh}(z+\tfrac{\mathrm{i}\gamma}{2}-s_\ell)}\right] \frac{q(z+\mathrm{i}\gamma)}{q(z-\mathrm{i}\gamma)} \quad (3.3)$$

associated with the unique solution $\{\lambda_\ell\}_{\ell=1}^{L/2} \equiv \{\lambda\}$, $1+\mathfrak{a}(\lambda_j)=0$. Obviously L has to be even and eigenvalue as well as auxiliary function are periodic in $\mathrm{i}\pi$. Because of this periodicity the boundary parameters $\xi^\pm$ can be restricted to the complex interval $(-\mathrm{i}\pi/2, \mathrm{i}\pi/2]$ for an hermitian Hamiltonian (1.1). Once a set of Bethe numbers $\{\lambda\} = \{\lambda_\ell\}_{\ell=1}^{L/2}$ is fixed satisfying $1+\mathfrak{a}(\lambda_j)=0$ for all $j=1,\ldots,L/2$ there are additional hole-type solutions $\{\chi\} = \{\chi_k\}_{k=1}^{L+1}$ to the same equation, $1+\mathfrak{a}(\chi_k)=0$. These are also zeros of the eigenvalue (3.2),

$$\Lambda(z) = \frac{(-1)^L \phi(z+\tfrac{\eta}{2}) q(z-\eta) \operatorname{sh}(2z+\eta)}{2\operatorname{ch}(z+\tfrac{\eta}{2})\operatorname{ch}(z-\tfrac{\eta}{2}) \operatorname{sh}(2z)} \\ \times \frac{\operatorname{sh}(z+\xi^+ - \tfrac{\eta}{2})\operatorname{sh}(z+\xi^- - \tfrac{\eta}{2})}{\operatorname{sh}\xi^+ \operatorname{sh}\xi^-} \frac{1+\mathfrak{a}(z)}{q(z)}. \quad (3.4)$$

The number of holes follows from the transformation $w \equiv \mathrm{e}^{2z}$ of $1+\mathfrak{a}(z)$ into a rational function of w where the nominator is a polynomial of degree $3L+4$: Due to the symmetry $\mathfrak{a}(-z) = 1/\mathfrak{a}(z)$ all zeros λ_j and χ_k appear twice with different signs and thus they are symmetrically distributed with respect to the origin as shown in figure 3.1. Additionally the equation $1+\mathfrak{a}(z)=0$ has two trivial solutions $z=0$ and $z=\mathrm{i}\pi/2$ fixing the number of

3. Scalar Products for the Diagonal XXZ Spin Chain

	$\frac{\pi}{2} \geq \frac{\xi^+}{i} > \gamma$	$\gamma \geq \frac{\xi^+}{i} > \frac{\gamma}{2}$	$\frac{\gamma}{2} \geq \frac{\xi^+}{i} > 0$	$0 > \frac{\xi^+}{i} > -\frac{\pi}{2}$
$\frac{\pi}{2} \geq \frac{\xi^-}{i} > \gamma$	I			
$\gamma \geq \frac{\xi^-}{i} > \frac{\gamma}{2}$	II	III		
$\frac{\gamma}{2} \geq \frac{\xi^-}{i} > 0$	IV	V	VI	
$0 > \frac{\xi^-}{i} > -\frac{\pi}{2}$	VII	VIII	IX	X

Table 3.1: Ten possible combinations of the boundary fields $\xi^\pm$.

hole-type solutions to be $L + 1$.

Compared to the case of the half infinite chain where the problem can be treated by root densities in the thermodynamic limit we want to pursue another way [21, 57, 58] valid for a finite number of lattice sites. It turns out that the meromorphic function $\mathfrak{a}(z)$ is sufficiently well determined by the gross properties of $\{\lambda\}$ and $\{\chi\}$ depending on the value of both boundary fields. As (3.3) is symmetric in the parameters $\xi^\pm$, one has to distinguish between ten main cases, c.f. table 3.1, for the pole structure in view of $\xi^\pm$, the positions of Bethe numbers and hole-type solutions. Indeed, the last case X is sensitive to the values of $\xi^\pm$ but inverting all parameters $\xi^\pm \to -\xi^\pm$ formally reverses the z-direction and maps the region X to the cases I through VI.

Let us consider some examples by looking at the first column of table 3.1. From numerical calculations one observes for opposite boundary fields (region VII) $L/2$ real Bethe roots inside the strip $|\operatorname{Im} z| < \gamma/2$ and finds the hole-type solutions to lie outside of it. Additionally one hole-type solution seems to stick to the pole $z = \eta/2 - \xi^-$ of the boundary field (figure 3.1, right panel). This observation along with the eigenvalue and the known asymptotics is enough to derive a set of equations relating the second logarithmic derivatives of $\mathfrak{a}$ and $(1 + \mathfrak{a})$ to each other determining $\mathfrak{a}(z)$ uniquely by means of the integral Fourier transform. Especially the exact position of the Bethe roots is not to enter the equations. To achieve this the q-functions in equation (3.3) have to be eliminated by introducing auxiliary functions $\mathfrak{b}(z) = (\mathfrak{a}(z - i\gamma/2))^{-1}$ and $\bar{\mathfrak{b}}(z) = \mathfrak{a}(z + i\gamma/2)$. Then the Fourier transform of the second logarithmic derivative of q can be expressed through a combination of the Fourier transforms of $(1 + \bar{\mathfrak{b}})$ and $1 + 1/\mathfrak{b}$. Resubstitution into the Fourier transforms of the auxiliary functions $\mathfrak{b}$ and $\bar{\mathfrak{b}}$ already yields the aspired connection. The details of this technique are explained

3.1. Auxiliary Function

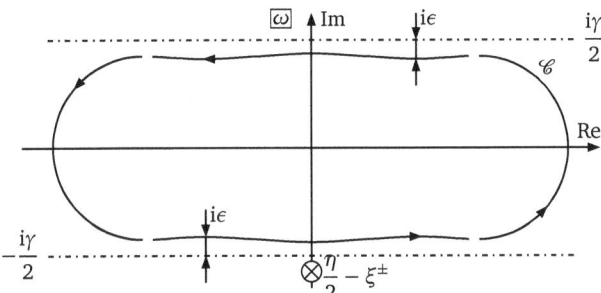

Figure 3.2: The Bethe roots are enclosed by the canonical contour $\mathscr{C}$ for the massless case in its parametrization $\eta = i\gamma$, $0 < \gamma < \pi/2$. Note the symmetry $\mathfrak{a}(-z) = 1/\mathfrak{a}(z)$ mirroring all solutions of the equation $1 + \mathfrak{a}(z) = 0$ at the origin. The poles at $\omega = \eta/2 - \xi^{\pm}$ of the function $(1+\mathfrak{a})$ have to lie outside the contour.

in [58] and we may leave with the homogeneous $s_j \to 0$ result

$$\ln \mathfrak{a}(z) = 4\eta + \ln \left[\frac{\sh(2z-\eta)\sh(z-\eta)}{\sh(2z+\eta)\sh(z+\eta)} \right]$$
$$- 2\eta + \ln \left[\frac{\sh(z-(\xi^+ - \frac{\eta}{2}))\sh(z-(\xi^- - \frac{\eta}{2}))}{\sh(z+(\xi^+ - \frac{\eta}{2}))\sh(z+(\xi^- - \frac{\eta}{2}))} \right] \qquad (3.5)$$
$$+ 2\eta L + 2L \ln \left[\frac{\sh(z-\frac{\eta}{2})}{\sh(z+\frac{\eta}{2})} \right] - \int_{\mathscr{C}} \frac{d\omega}{2\pi i} \frac{\sh(2\eta)\ln(1+\mathfrak{a}(\omega))}{\sh(z-\omega+\eta)\sh(z-\omega-\eta)}$$

valid for the region $|\Im z| \leq \gamma/2 - \varepsilon$. The factor $\varepsilon \ll 1$ ensures the Fourier integrals to converge and serves in the inhomogeneous case as a convenient restriction $|\Im s_j| < \varepsilon$. The canonical contour $\mathscr{C}$ is depicted in figure 3.2 and extends to infinity. Compared to the periodic chain [21] only the first and second lines of driving terms were added.

For boundary fields exceeding $i\gamma$ (region I) a hole-type solution on the real line appears besides the outermost Bethe root (figure 3.1, left panel). This is due to a change in the description of the ground state encoded by $L/2$ real Bethe roots. All other L hole-type solutions remain outside the strip $|\Im z| < \gamma/2$. The hole-type solution on the real line corresponds to the term

$$+4\eta + \ln \left[\frac{\sh(z+\chi-\eta)\sh(z-\chi-\eta)}{\sh(z+\chi+\eta)\sh(z-\chi+\eta)} \right] \qquad (3.6)$$

which has to be added on the RHS of (3.5) due to complex analysis imposing the additional constraint $1 + \mathfrak{a}(\chi) = 0$ on the auxiliary function. This is similar to the case of excited states in the periodic XXZ chain [59].

For boundary fields $\gamma - \varepsilon > \xi^-/i > 0$ (regions II, IV) the pole at $z = \eta/2 - \xi^-$ has to remain outside the contour guaranteed by a deformation. Applying the residue theorem

3. Scalar Products for the Diagonal XXZ Spin Chain

	I	II	III	IV	V	VI	VII	VIII	IX
hole-type solution	•	•	•	•	•	•			
left boundary pole (−)	•	•	•	•	•				
right boundary pole (+)		•		•	•			•	•

Table 3.2: Driving terms for the possible combinations of boundary fields $\xi^\pm$

yields the additional driving term

$$-2\eta + \ln\left[\frac{\operatorname{sh}(z + \eta/2 + \xi^-)}{\operatorname{sh}(z - 3\eta/2 + \xi^-)}\right] \tag{3.7}$$

which, besides (3.6), has to be added on the RHS of (3.5) for that case. This is the only modification compared to case I because the structure of the root distribution of λ_j and χ_k with respect to the strip $|\operatorname{Im} z| < \gamma/2$ is unchanged. Nevertheless approaching zero from below with the pole $z = \eta/2 - \xi^-$ the Bethe root closest to the origin moves towards zero along the real axis (region II). Passing the origin the pole picks up this Bethe root and pulls it (up to exponential corrections with respect to the chain length) along the positive imaginary axis until the upper part of the canonical contour is reached. Because of this imaginary Bethe root the corresponding states are termed *boundary bound states* (region IV) [97].

Especially in the *XXX* limit when the parallely oriented boundary fields (region VI, two imaginary Bethe roots) become strong enough to significantly arrest the outermost spins of the chain the considered *boundary bound state* with $M = L/2$ Bethe roots and a total magnetization of zero refers no longer to the ground state. Here the almost fixed boundary spins can be regarded as effective boundary fields for a spin chain with two lattice sites and one Bethe root less. For the *XXZ* chain this effect already sets in for the regions V, VI, VIII, IX but depends on the values of the anisotropy $\operatorname{ch}\eta$ compared to the boundary fields.

Numerics suggest that all cases I to IX with zero magnetization have $L/2$ Bethe numbers within the contour $\mathscr{C}$ as in the considered examples above (see appendix B). Thus all possible forms of driving terms with respect to hole-type solutions and poles of the boundary fields are given. The non-linear integral equation for the auxiliary function can then be fixed if one considers table 3.2 as a building block. Here • marks the additional driving terms (3.6) and (3.7) which have to be added on the RHS of (3.5) for each single case. This accounts for the hole-type solution χ inside the canonical contour imposing $1 + \mathfrak{a}(\chi) = 0$ and the boundary fields $\xi^\pm$.

In the following two sections we shall derive our main result of this chapter valid for distributions of Bethe numbers in the strip $|\operatorname{Im} z| < \gamma/2$ according to the left panel of figure 3.1. For this reason, we have to introduce a closed contour $\mathscr{C}'$ similar to $\mathscr{C}$ but excluding all hole-type solutions, especially the one on the real line closing the set of Bethe numbers if present.

Remark. *As in the case of the periodic XXZ chain it is possible to derive a non-linear integral equation for the eigenvalue in terms of the auxiliary function [58]. However the calculations are tedious and the eigenvalue (3.2) can be determined from the Bethe-equations (3.1) directly. Hence it is more motivating to pursue the way of calculating correlation functions.*

3.2 Integral Representation for the Determinant Formula

To calculate scalar-valued expectation values of local operators a nice combinatorial result for the Bethe-eigenvectors of the open XXZ chain applies. We will use the choice of shifting the operator content to the right reflection algebra (2.45) within this chapter and hence work with $\prod_{b=1}^{M} \mathscr{B}^{(+)}(\lambda_b)|0\rangle$ as the Bethe-eigenvector.

First note that it is possible to reduce the $\mathscr{B}^{(+)}$ operator generating Bethe states by applying (2.46) to operators of the periodic chain

$$\mathscr{B}^{(+)}(\lambda) = \frac{1}{2\operatorname{ch}(\lambda + \frac{\eta}{2})\operatorname{sh}\xi^+} \frac{\operatorname{sh}(2\lambda + \eta)}{\operatorname{sh}(2\lambda)}$$
$$\times \Big[\operatorname{sh}(\lambda - \tfrac{\eta}{2} + \xi^+) B(\lambda - \tfrac{\eta}{2}) D(-\lambda - \tfrac{\eta}{2})$$
$$+ \operatorname{sh}(\lambda + \tfrac{\eta}{2} - \xi^+) B(-\lambda - \tfrac{\eta}{2}) D(\lambda - \tfrac{\eta}{2}) \Big]. \tag{3.8}$$

Then the key element is the inversion formula

$$e_{m\,\alpha}^{\ \beta} = \Big[\prod_{j=1}^{m-1}\big(A(s_j) + D(s_j)\big)\Big] T^{\beta}_{\alpha}(s_m) \Big[\prod_{j=1}^{m}\big(A(s_j) + D(s_j)\big)^{-1}\Big] \tag{3.9}$$

for the standard basis $(e_{\alpha}^{\ \beta})^{\alpha'}_{\ \beta'} = \delta^{\alpha'}_{\alpha}\delta^{\beta}_{\beta'}$ at site m from [56]. Because it is written in terms of entries of the monodromy $T(\lambda)$ the action on a Bethe state with (3.8) can be computed by simply applying the Yang-Baxter algebra. Then some Bethe numbers λ_j are replaced [54] by pairwise distinct lattice inhomogeneities $\zeta_k = \eta/2 + s_k$ to regularize the expressions

$$(e_{m\,\alpha}^{\ \beta})\Big[\prod_{b=1}^{M}\mathscr{B}^{(+)}(\lambda_b)\Big]|0\rangle = \sum_{\alpha_m} C_{\alpha_m}\big(\{\lambda_j\}_{j=1}^{M}, \{\zeta_k\}_{k=1}^{m}\big)\Big[\prod_{b\in\alpha_m}\mathscr{B}^{(+)}(\mu_b)\Big]|0\rangle. \tag{3.10}$$

Here we have $\{\mu_b\} = \{\lambda_j\}_{j=1}^{M} \cup \{\zeta_k\}_{k=1}^{m}$ and the summation is taken over certain subsets α_m of $\{1, 2, \ldots, M+m\}$. For a local operator at site m only the first m inhomogeneities $s_1, \ldots, s_m$ enter and their shift of $\eta/2$, $\zeta_k = s_k + \eta/2$, is due to the explicit decomposition[1] of $\mathscr{B}^{(+)}(\lambda)$

---

[1] Note that the operator $\mathscr{B}^{(+)}(\lambda)$ here, (3.8), and the corresponding expression in Kitanine *et al.* differs by an overall prefactor and a shift of $\eta/2$ in the periodic chain operators. Looking up [54] we find

$$\mathscr{B}^{(+)}_{\text{Kitanine}}(\lambda) = (-1)^L \frac{\operatorname{sh}(2\lambda + \eta)}{\operatorname{sh}(2\lambda)} \Big[\operatorname{sh}(\lambda - \tfrac{\eta}{2} + \xi^+) B(\lambda) D(-\lambda) + \operatorname{sh}(\lambda + \tfrac{\eta}{2} - \xi^+) B(-\lambda) D(\lambda) \Big].$$

To make use of the normalized scalar product formula (3.11) one should always bear in mind, that the right Bethe vector containing $\mathscr{B}^{(+)}(\mu_k)$ gets its arguments from commutations starting with $\mathscr{B}^{(+)}(\lambda_j)$ such that the prefactors in front of the square brackets cancel due to the normalization.

3. Scalar Products for the Diagonal XXZ Spin Chain

in terms of the periodic chain operators. The coefficients C_{α_m} can be computed generically and for an illustrating example to this formula see (3.28).

Proposition [54]. *For a set of pairwise distinct numbers $\{\mu_k\}_{k=1}^M$ and Bethe roots $\{\lambda_l\}_{l=1}^M$ solving the Bethe ansatz equations (3.1) the normalized determinant formula for scalar products including members of the right reflection algebra reads*

$$\frac{\langle 0 \| [\prod_{a=1}^M \mathscr{C}^{(+)}(\lambda_a)] [\prod_{b=1}^M \mathscr{B}^{(+)}(\mu_b)] |0\rangle}{\langle 0 \| [\prod_{a=1}^M \mathscr{C}^{(+)}(\lambda_a)] [\prod_{b=1}^M \mathscr{B}^{(+)}(\lambda_b)] |0\rangle} = \tag{3.11}$$

$$\left[\prod_{a<b} \frac{\operatorname{sh}(\lambda_{ab})\operatorname{sh}(\overline{\lambda_{ab}})}{\operatorname{sh}(\mu_{ab})\operatorname{sh}(\overline{\mu_{ab}})}\right] \left[\prod_{l=1}^M \frac{\operatorname{sh}(2\mu_l + \eta)\operatorname{sh}(2\lambda_l)}{\operatorname{sh}(2\lambda_l + \eta)\operatorname{sh}(2\mu_l)}\right] \frac{\det\left[H(\lambda_j, \mu_k)\right]_{j,k=1,\ldots,M}}{\det\left[H(\lambda_j, \lambda_k)\right]_{j,k=1,\ldots,M}}$$

with the entry

$$H(\lambda_j, \mu_k) \equiv \frac{y_j(\mu_k) - y_j(-\mu_k)}{\operatorname{sh}(\lambda_j - \mu_k)\operatorname{sh}(\lambda_j + \mu_k)} \tag{3.12}$$

of the determinant, the shorthands $\lambda_{ab} \equiv \lambda_a - \lambda_b$, $\overline{\lambda_{ab}} \equiv \lambda_a + \lambda_b$ and the set $\{\lambda\}$ of Bethe roots included in the functions

$$y_j(z) = \frac{\hat{y}(z, \{\lambda\})}{\operatorname{sh}(z - \lambda_j - \eta)\operatorname{sh}(z + \lambda_j - \eta)} \tag{3.13}$$

$$\hat{y}(z, \{\lambda\}) \equiv \mathfrak{a}(z - \tfrac{\eta}{2})\mathfrak{d}(-z - \tfrac{\eta}{2})\operatorname{sh}(z + \xi^+ - \tfrac{\eta}{2})\operatorname{sh}(z + \xi^- - \tfrac{\eta}{2})$$

$$\times \left[\prod_{l=1}^M \operatorname{sh}(z - \lambda_l - \eta)\operatorname{sh}(z + \lambda_l - \eta)\right]. \tag{3.14}$$

Here $\mathfrak{a}(\lambda) = \prod_{\ell=1}^L \operatorname{sh}(\lambda - s_\ell + \eta)$ and $\mathfrak{d}(\lambda) = \prod_{\ell=1}^L \operatorname{sh}(\lambda - s_\ell)$ are the vacuum expectation values of the operators $A(\lambda)$ and $D(\lambda)$ of the periodic chain with inhomogeneities s_l approaching zero in the homogeneous limit.

The Bethe ansatz equations (3.1) follow from

$$\frac{y_j(-z)}{y_j(z)} = \mathfrak{a}(z)\frac{\operatorname{sh}(2z + \eta)}{\operatorname{sh}(2z - \eta)}\frac{\operatorname{sh}(z + \lambda_j - \eta)\operatorname{sh}(z - \lambda_j - \eta)}{\operatorname{sh}(z + \lambda_j + \eta)\operatorname{sh}(z - \lambda_j + \eta)} \tag{3.15}$$

and can be rewritten as $y_j(\lambda_j) = y_j(-\lambda_j)$, $j = 1,\ldots,M$ allowing to recast the entries of the determinant in the form

$$H(\lambda_j, \mu_k) = \frac{y_j(\mu_k)\operatorname{sh}(\mu_k + \lambda_j - \eta)\operatorname{sh}(\mu_k - \lambda_j - \eta)}{\operatorname{sh}(2\mu_k - \eta)\operatorname{sh}(\lambda_j - \mu_k)\operatorname{sh}(\lambda_j + \mu_k)}$$

$$\times \left\{\frac{\operatorname{sh}(2\mu_k - \eta)}{\operatorname{sh}(\mu_k + \lambda_j - \eta)\operatorname{sh}(\mu_k - \lambda_j - \eta)} - \mathfrak{a}(\mu_k)\frac{\operatorname{sh}(2\mu_k + \eta)}{\operatorname{sh}(\mu_k + \lambda_j + \eta)\operatorname{sh}(\mu_k - \lambda_j + \eta)}\right\}. \tag{3.16}$$

3.2. Integral Representation for the Determinant Formula

Considering the limit $\mu_k \to \lambda_k$ in the above expression to get in contact with the desired matrix elements yields

$$\lim_{\mu_k \to \lambda_k} \frac{1}{\operatorname{sh}(\lambda_j + \mu_k)\operatorname{sh}(\lambda_j - \mu_k)} \left\{ \frac{\operatorname{sh}(2\mu_k - \eta)}{\operatorname{sh}(\mu_k + \lambda_j - \eta)\operatorname{sh}(\mu_k - \lambda_j - \eta)} \right.$$
$$\left. - \mathfrak{a}(\mu_k) \frac{\operatorname{sh}(2\mu_k + \eta)}{\operatorname{sh}(\mu_k + \lambda_j + \eta)\operatorname{sh}(\mu_k - \lambda_j + \eta)} \right\} \quad (3.17)$$
$$= \frac{1}{\operatorname{sh}\eta \operatorname{sh}(2\lambda_j)} \left[iK_\eta(\lambda_j + \lambda_k) - iK_\eta(\lambda_j - \lambda_k) - \delta_k^j \frac{\partial \ln \mathfrak{a}}{\partial z}(\lambda_j) \right]$$

where one separately has to treat the case $\mu_k \to \lambda_j$ accounting for the Kronecker δ_k^j by virtue of l'Hospital's rule. The kernel K_η from the auxiliary function is

$$K_\eta(\lambda) = \frac{1}{i} \frac{\operatorname{sh}(2\eta)}{\operatorname{sh}(\lambda + \eta)\operatorname{sh}(\lambda - \eta)}. \quad (3.18)$$

Obviously all normalized expectation values (3.11) contain the elementary ratio

$$\frac{\det\left[\psi(\lambda_a, \mu_b)\right]_{a,b=1,\dots,M}}{\det\left[\phi(\lambda_j, \lambda_k)\right]_{j,k=1,\dots,M}} = \det\left[\phi^{-1}(\lambda_j, \lambda_k)\psi(\lambda_k, \mu_\ell)\right]_{j,\ell=1,\dots,M} \quad (3.19)$$

where on the right hand side $\phi^{-1}(\lambda_j, \lambda_k)$ denote the entries of the inverse matrix and summation over k is understood. Now for reshaping the right hand side we closely follow [36] as similar computations were done for the periodic XXZ chain. Here the entries are

$$\phi(\lambda_j, \lambda_k) = \frac{1}{\operatorname{sh}(2\lambda_j)} \left[iK_\eta(\lambda_j + \lambda_k) - iK_\eta(\lambda_j - \lambda_k) - \delta_k^j \frac{\partial \ln \mathfrak{a}}{\partial z}(\lambda_j) \right] \quad (3.20)$$

$$\psi(\lambda_j, \mu_k) = \frac{\operatorname{sh}\eta \operatorname{sh}(2\mu_k - \eta)}{\operatorname{sh}(\lambda_j - \mu_k)\operatorname{sh}(\mu_k - \lambda_j - \eta)\operatorname{sh}(\mu_k + \lambda_j - \eta)\operatorname{sh}(\lambda_j + \mu_k)}$$
$$- \mathfrak{a}(\mu_k) \frac{\operatorname{sh}\eta \operatorname{sh}(2\mu_k + \eta)}{\operatorname{sh}(\lambda_j - \mu_k)\operatorname{sh}(\mu_k - \lambda_j + \eta)\operatorname{sh}(\mu_k + \lambda_j + \eta)\operatorname{sh}(\lambda_j + \mu_k)} \quad (3.21)$$

defining the new matrix $J(\lambda_j, \mu_l) \equiv \phi^{-1}(\lambda_j, \lambda_k)\psi(\lambda_k, \mu_l)$. How to express the determinant (3.19) in terms of a density function is presented in appendix A and summarized in the following lemma.

Lemma 1. *For simplicity assume $\{\mu_l\}_{l=1}^M$ to be a copy of the Bethe numbers where the first n roots $\lambda_1, \dots, \lambda_n$ are replaced by some c-numbers $\nu_1, \dots, \nu_n$ considered as lattice inhomogeneities $\zeta_j = \eta/2 + s_j$ from the strip $|\operatorname{Im}(\zeta_j - \eta/2)| < \varepsilon$. Then the determinant from above reduces to*

$$\frac{\det\left[\psi(\lambda_a, \mu_b)\right]_{a,b=1,\dots,M}}{\det\left[\phi(\lambda_j, \lambda_k)\right]_{j,k=1,\dots,M}} = \det\left[J(\lambda_j, \nu_\ell)\right]_{j,\ell=1,\dots,n} = \det\left[\frac{G(\lambda_j, \nu_\ell)}{\mathfrak{a}'(\lambda_j)}\right]_{j,\ell=1,\dots,n} \quad (3.22)$$

3. Scalar Products for the Diagonal XXZ Spin Chain

where $'$ denotes a derivative and $G(\lambda, v)$ is the solution to the linear integral equation

$$G(\lambda, v) = \frac{\operatorname{sh}\eta}{\operatorname{sh}(\lambda+v)\operatorname{sh}(\lambda+v-\eta)} - \frac{\operatorname{sh}\eta}{\operatorname{sh}(\lambda-v)\operatorname{sh}(\lambda-v+\eta)}$$
$$+ \int_{\mathscr{C}'} \frac{\mathrm{d}\omega}{2\pi\mathrm{i}} \frac{\operatorname{sh}(2\eta)}{\operatorname{sh}(\lambda-\omega+\eta)\operatorname{sh}(\lambda-\omega-\eta)} \frac{G(\omega, v)}{1+\mathfrak{a}(\omega)} \quad (3.23)$$

on the contour $\mathscr{C}'$. Here we already made use of $\mathfrak{a}(\zeta_j) = 0$ and minded the simple zeros $G(0, v) = G(\lambda, \eta/2) = 0$. The density function shows the symmetry $G(-\lambda, v) = -G(\lambda, v)$ with respect to the first argument λ whereas here the second argument v is restricted to the strip $|\operatorname{Im}(v - \eta/2)| < \varepsilon$ outside $\mathscr{C}'$. The contour $\mathscr{C}'$ excludes the hole-type solutions χ and depends on the parameter ε as shown in figure 3.3.

To generalize the result let us introduce the disjoint union of the sets $\{\lambda\} = \{\lambda^+\} \cup \{\lambda^-\}$ and $\{\mu\} = \{\mu^+\} \cup \{\lambda^-\}$ and denote the cardinality of the partitions $\{\lambda^\pm\}$ by $|\lambda^\pm|$. Then along with the slightly modified function [54]

$$\mathscr{S}_\sigma(\{\lambda^+\}, \{\mu^+\}|\{\lambda^-\}) = \left[\prod_{a=1}^n \frac{\widehat{y}(\mu_a^+, \{\lambda\})\operatorname{sh}(2\mu_a^+ + \eta)}{\operatorname{sh}(2\mu_a^+)\operatorname{sh}(2\mu_a^+ - \eta)} \frac{\operatorname{sh}(2\lambda_a^+)\operatorname{sh}(2\sigma_a^+\lambda_a^+ - \eta)}{\widehat{y}(\sigma_a^+\lambda_a^+, \{\lambda\})\operatorname{sh}(2\lambda_a^+ + \eta)}\right]$$
$$\times \left[\prod_{a<b} \frac{\operatorname{sh}(\lambda_a^+ - \lambda_b^+)\operatorname{sh}(\lambda_a^+ + \lambda_b^+)}{\operatorname{sh}(\mu_a^+ - \mu_b^+)\operatorname{sh}(\mu_a^+ + \mu_b^+)}\right] \left[\prod_{a=1}^n \prod_{b=1}^{M-n} \frac{\operatorname{sh}(\lambda_a^+ - \lambda_b^-)\operatorname{sh}(\lambda_a^+ + \lambda_b^-)}{\operatorname{sh}(\mu_a^+ - \lambda_b^-)\operatorname{sh}(\mu_a^+ + \lambda_b^-)}\right] \quad (3.24)$$

the normalized scalar product

$$\frac{\langle 0 | \left[\prod_{a=1}^M \mathscr{C}^{(+)}(\lambda_a)\right] \left[\prod_{b=1}^M \mathscr{B}^{(+)}(\mu_b)\right] | 0 \rangle}{\langle 0 | \left[\prod_{a=1}^M \mathscr{C}^{(+)}(\lambda_a)\right] \left[\prod_{b=1}^M \mathscr{B}^{(+)}(\lambda_b)\right] | 0 \rangle}$$
$$= \mathscr{S}_\sigma(\{\lambda^+\}, \{\mu^+\}|\{\lambda^-\}) \det\left[\frac{G(\lambda_j^+, \mu_k^+)}{\mathfrak{a}'(\lambda_j^+)}\right]_{j,k=1,\ldots,n} \quad (3.25)$$

effectively reduces with $|\lambda^+| = |\mu^+| = n$ to an $n \times n$ matrix. The set $\{\sigma\}$ with $\sigma_j = \pm 1$ accounts for the symmetry of the Bethe roots which can be seen from the Bethe ansatz equations in the form $\widehat{y}(\lambda_j, \{\lambda\})\operatorname{sh}(-2\lambda_j - \eta) = \widehat{y}(-\lambda_j, \{\lambda\})\operatorname{sh}(2\lambda_j - \eta)$ for $j = 1, \ldots, M$ leaving $\mathscr{S}_\sigma$ unchanged.

3.3 Generating Function of the Magnetization

For an illustrating example we shall now apply the integral representation of the scalar product formula (3.25) to a generating function of the S^z-magnetization. Note that we will assume the case of one hole-type solution on the real line accounting for $0 < \xi^\pm/\mathrm{i} < \pi/2$.

Proposition [55]. *Corresponding to one of the simplest non-trivial one-point functions in the open spin chain is the one-parameter generating function*

$$Q_m(\varphi) = \left[\prod_{j=1}^m \left(A(s_j) + e^\varphi D(s_j)\right)\right] \left[\prod_{j=1}^m \left(A(s_j) + D(s_j)\right)^{-1}\right] \quad (3.26)$$

3.3. Generating Function of the Magnetization

of the longitudinal magnetization

$$\left\langle \frac{1-\sigma_m^z}{2} \right\rangle = D_m \partial_\varphi \langle Q_m(\varphi) \rangle \Big|_{\varphi=0} . \tag{3.27}$$

It includes a discrete derivative $D_m u_m = u_m - u_{m-1}$ on the lattice and a continuous one with respect to φ. Its action on a Bethe state reads

$$Q_m(\varphi)\left[\prod_{\ell=1}^{M} \mathcal{B}^{(+)}(\lambda_\ell)\right]|0\rangle = \sum_{n=0}^{m} \sum_{|\lambda^+|=n} \sum_{|\zeta^+|=n} \left[\prod_{j=1}^{n} \sum_{\sigma_j=\pm 1}\right] \det\left[M(\sigma_j^+ \lambda_j^+, \zeta_k^+)\right]_{j,k=1,\ldots,n}$$

$$\times \left[\prod_{a=1}^{n} \sigma_a^+\right] W_-(\{\sigma^+ \lambda^+\}, \{\zeta^+\}) \left[\prod_{a=1}^{n} \frac{b(\sigma_a^+ \lambda_a^+)}{b'(\zeta_a^+)} \frac{1}{\text{sh}(2\zeta_a^+ - \eta)}\right] \tag{3.28}$$

$$\times \mathcal{S}_\sigma^{-1}(\{\lambda^+\}, \{\zeta^+\}|\{\lambda^-\}) \left[\prod_{a=1}^{n} \mathcal{B}^{(+)}(\zeta_a)\right] \left[\prod_{b=1}^{M-n} \mathcal{B}^{(+)}(\lambda_b^-)\right]|0\rangle$$

with the known matrix

$$M(\lambda_j, \mu_k) = \frac{\text{sh}\,\eta}{\text{sh}(\lambda_j - \mu_k)\,\text{sh}(\lambda_j - \mu_k - \eta)} +$$

$$+ \frac{e^\varphi \text{sh}\,\eta}{\text{sh}(\lambda_j - \mu_k)\,\text{sh}(\lambda_j - \mu_k + \eta)} \left[\prod_{\ell=1}^{n} \frac{\text{sh}(\lambda_j - \lambda_\ell^+ - \eta)\,\text{sh}(\lambda_j - \mu_\ell^+ + \eta)}{\text{sh}(\lambda_j - \mu_\ell^+ - \eta)\,\text{sh}(\lambda_j - \lambda_\ell^+ + \eta)}\right] \tag{3.29}$$

from the generating function of the zz-correlation and the function

$$\frac{W_-(\{\omega\}, \{z\})}{W(\{\omega\}, \{z\})} = \left[\prod_{\ell=1}^{n} \frac{\text{sh}(z_\ell + \xi^- - \frac{\eta}{2})}{\text{sh}(\omega_\ell + \xi^- - \frac{\eta}{2})}\right] \left[\frac{\prod_{a,b=1}^{n} \text{sh}(z_b + \omega_a - \eta)}{\prod_{a<b} \text{sh}(z_a + z_b - \eta)\,\text{sh}(\omega_a + \omega_b - \eta)}\right] \tag{3.30}$$

$$W(\{\omega\}, \{z\}) = \left[\prod_{a,b=1}^{n} \frac{\text{sh}(z_b - \omega_a - \eta)\,\text{sh}(z_b - \omega_a + \eta)}{\text{sh}(\omega_a - \omega_b - \eta)\,\text{sh}(z_a - z_b + \eta)}\right] \tag{3.31}$$

picking out the left boundary with ξ^- to start counting the lattice sites. All inhomogeneities entering the generating function are included within the expressions [36]

$$b(\lambda) = \left[\prod_{\ell=1}^{m} \frac{\text{sh}(\lambda - \zeta_\ell)}{\text{sh}(\lambda - \zeta_\ell + \eta)}\right] \quad , \quad \frac{1}{b'(\zeta_j)} = \frac{\prod_{\ell=1}^{m} \text{sh}(\zeta_j - \zeta_\ell + \eta)}{\prod_{\substack{\ell=1 \\ l \neq j}}^{m} \text{sh}(\zeta_j - \zeta_\ell)} . \tag{3.32}$$

The function $\mathcal{S}_\sigma$ already appeared in the scalar product formula (3.25), whereas $\sigma_j = \pm 1$ accounts for the symmetry of the Bethe roots.

Note that here the combinatorial part is expressed by the set of all ordered pairs $(\{\lambda^+\}, \{\lambda^-\})$ of fixed cardinality $|\lambda^+|$ and $|\zeta^+|$ respectively indexing the sums. Switching to the normalized scalar product the expectation value of the generating function can be written in terms

3. Scalar Products for the Diagonal XXZ Spin Chain

of the density function $G(\sigma_j \lambda_j, \nu) = \sigma_j G(\lambda_j, \nu)$,

$$\langle Q_m(\varphi) \rangle = \frac{\langle 0 \| [\prod_{a=1}^{M} \mathscr{C}^{(+)}(\lambda_a)] Q_m(\varphi) [\prod_{b=1}^{M} \mathscr{B}^{(+)}(\lambda_b)] | 0 \rangle}{\langle 0 \| [\prod_{a=1}^{M} \mathscr{C}^{(+)}(\lambda_a)] [\prod_{b=1}^{M} \mathscr{B}^{(+)}(\lambda_b)] | 0 \rangle}$$

$$= \sum_{n=0}^{m} \sum_{|\lambda^+|=n} \sum_{|\zeta^+|=n} \left[\prod_{j=1}^{n} \sum_{\sigma_j=\pm 1} \right] \left[\prod_{a=1}^{n} \frac{\mathfrak{b}(\sigma_a^+ \lambda_a^+)}{\mathfrak{b}'(\zeta_a^+)} \frac{1}{\operatorname{sh}(2\zeta_a^+ - \eta)} \right] \quad (3.33)$$

$$\times W_-(\{\sigma^+ \lambda^+\}, \{\zeta^+\}) \det \left[M(\sigma_j^+ \lambda_j^+, \zeta_k^+) \right] \det \left[\frac{G(\sigma_j^+ \lambda_j^+, \zeta_k^+)}{\mathfrak{a}'(\lambda_j^+)} \right]$$

with indices of the determinant $j, k = 1, \ldots, n$.

The last step now is to get rid of the explicit dependence on Bethe roots by integrals according to the following lemma.

Lemma 2. *Let $f(\omega_1, \ldots, \omega_n)$ be a complex function, symmetric in its arguments and equal to zero if any two of its arguments agree up to a sign. Furthermore if it is analytic on and inside the simple n-fold contour $(\mathscr{C}')^n$ and shows a simple zero at $\omega_j = 0$ to compensate the first order pole of the auxiliary function $1/(1 + \mathfrak{a})$ then*

$$\sum_{|\lambda^+|=n} \sum_{\sigma_1^+=\pm 1} \cdots \sum_{\sigma_n^+=\pm 1} \frac{f(\sigma_1^+ \lambda_1^+, \ldots, \sigma_n^+ \lambda_n^+)}{\prod_{\ell=1}^{n} \mathfrak{a}'(\lambda_\ell^+)}$$

$$= \frac{1}{n!} \left[\prod_{\ell=1}^{n} \int_{\mathscr{C}'} \frac{d\omega_\ell}{2\pi i} \frac{1}{1 + \mathfrak{a}(\omega_\ell)} \right] f(\omega_1, \ldots, \omega_n). \quad (3.34)$$

For lattice inhomogeneities within the strip $|\operatorname{Im}(\zeta_k - \eta/2)| < \varepsilon$ all poles with respect to the variable ω_j of the density function $G(\omega_j, \zeta_k)$ lie outside the contour $\mathscr{C}'$. The singularity of the function $W_-(\{\omega\}, \{\zeta\})$ at $\omega_j = \eta/2 - \xi^-$ is 'balanced' by the simple zero of $1/(1 + \mathfrak{a}(\omega_j))$ and because of the density function $G(0, \nu) = 0$ the expression in (3.33) meets – along with the determinant property – the conditions of lemma 2.

However, the same technique can be applied to the ζ^+-summation with inhomogeneities ζ_k in the vicinity of $\eta/2$ and thus outside of $\mathscr{C}'$. For a function f with the properties from above except for the simple zero at $\omega_j = 0$ the corresponding integrals read [36]

$$\sum_{|\xi^+|=n} \frac{f(\zeta_1^+, \ldots, \zeta_n^+)}{\prod_{\ell=1}^{n} \mathfrak{b}'(\zeta_\ell^+)} = \frac{1}{n!} \left[\prod_{\ell=1}^{n} \int_{\Gamma} \frac{dz_\ell}{2\pi i} \frac{1}{\mathfrak{b}(z_\ell)} \right] f(z_1, \ldots, z_n). \quad (3.35)$$

With respect to the integrand the contour Γ lies in the strip $|\operatorname{Im}(z - \eta/2)| < \varepsilon$ and surrounds all inhomogeneities $\zeta_1, \ldots, \zeta_m$. In addition due to the simple zero $G(\lambda, \eta/2) = 0$ the point $\eta/2$ can be enclosed enabling the homogeneous limit $\zeta_k \to \eta/2$ yielding as a main result the following proposition.

Proposition. *Let m be a site index counted from the left boundary and consider the functions $\mathfrak{b}(\lambda)$, $W_-(\{\omega\}, \{z\})$, $M(\omega_j, z_k)$ and $G(\omega_j, z_k)$ according to (3.32), (3.30), (3.29) and (3.23).*

3.3. Generating Function of the Magnetization

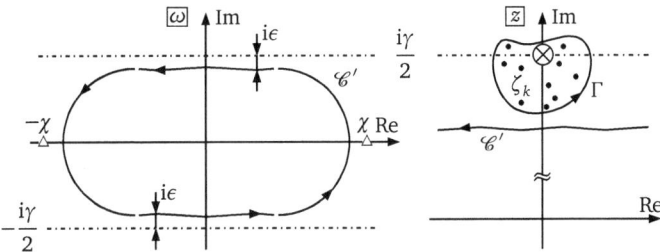

Figure 3.3: In the massless case $\eta = i\gamma$, $0 < \gamma < \pi/2$ the contour $\mathscr{C}'$ is limited by the hole-type solution χ as depicted in the left panel for a small $\varepsilon \ll 1$. The lattice inhomogeneities ζ_k lie in the vicinity of $\eta/2$ outside $\mathscr{C}'$ and are counterclockwisely surrounded by Γ (right panel).

Then the multiple integral representation of the generating function reads

$$\langle Q_m(\varphi)\rangle = \sum_{n=0}^{m} \frac{1}{(n!)^2} \left[\prod_{\ell=1}^{n} \int_{\mathscr{C}'} \frac{d\omega_\ell}{2\pi i} \frac{b(\omega_\ell)}{1+\mathfrak{a}(\omega_\ell)} \int_\Gamma \frac{dz_\ell}{2\pi i} \frac{1}{b(z_\ell)}\right] W_-(\{\omega\},\{z\})$$
$$\times \det\left[M(\omega_j, z_k)\right]_{j,k=1,\dots,n} \det\left[\frac{G(\omega_j, z_k)}{\operatorname{sh}(2z_k-\eta)}\right]_{j,k=1,\dots,n}. \tag{3.36}$$

The contours for the massless case are depicted in figure 3.3 and in the homogeneous limit the auxiliary function $\mathfrak{a}(z)$ is determined by the non-linear integral equation (3.5).

Thermodynamic Limit

Rewriting the density function (3.23) as the sum $G(\lambda, \nu) = G^+(\lambda, \nu) - G^+(\lambda, \eta - \nu)$ the partial density G^+ satisfies

$$G^+(\lambda, \nu') = -\frac{\operatorname{sh}\eta}{\operatorname{sh}(\lambda-\nu')\operatorname{sh}(\lambda-\nu'+\eta)} + \int_{\mathscr{C}'} \frac{d\omega}{2\pi i} \frac{\operatorname{sh}(2\eta)}{\operatorname{sh}(\lambda-\omega+\eta)\operatorname{sh}(\lambda-\omega-\eta)} \frac{G^+(\omega, \nu')}{1+\mathfrak{a}(\omega)} \tag{3.37}$$

for $|\operatorname{Im}(\nu'-\eta/2)| < \varepsilon$.

In the limit of infinitely many lattice sites i.e. $L \to \infty$, thus $\chi \to \infty$ and $\mathscr{C}' \to \mathscr{C}$, the auxiliary function $\mathfrak{a}$ is dominating for $\gamma < 0$ the upper part of the contour $\mathscr{C}$,

$$\ln \mathfrak{a}(\lambda) \sim 2L \frac{\operatorname{sh}(\lambda - \frac{i\gamma}{2})}{\operatorname{sh}(\lambda + \frac{i\gamma}{2})}, \tag{3.38}$$

such that only the lower part remains (figure 3.4, left panel). Clearly, the validity range of the variable ν' extends according to the pole structure of the driving term and in this limit

3. Scalar Products for the Diagonal XXZ Spin Chain

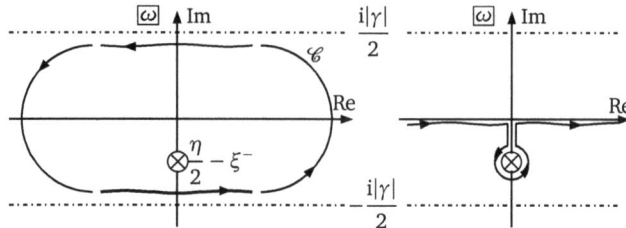

Figure 3.4: In the thermodynamic limit only the lower part of the contour $\mathscr{C}$ remains (left panel). Moving it towards the real axis the poles at $\omega_j = \eta/2 - \xi^-$ of the function W_- must not be crossed (right panel).

the density $G^+(\lambda, \nu') \to i\pi \rho(\lambda, \nu')$ satisfies

$$\rho(\lambda, \nu') + \int_{-\infty}^{\infty} \frac{d\omega}{2\pi} \frac{i\,\text{sh}(2\eta)\rho(\omega, \nu')}{\text{sh}(\lambda - \omega + \eta)\text{sh}(\lambda - \omega - \eta)} = \frac{i}{\pi} \frac{\text{sh}\,\eta}{\text{sh}(\lambda - \nu')\text{sh}(\lambda - \nu' + \eta)}. \quad (3.39)$$

Note that the variable ν' takes the values of ν and $\eta - \nu$ where ν is located in the vicinity of $\eta/2$. Applying the thermodynamic limit to the generating function one directly ends up with the result of Kitanine et al. [55] where $\gamma < 0$ is assumed,

$$\langle Q_m(\varphi) \rangle = \sum_{n=0}^{m} \frac{1}{(n!)^2} \left[\prod_{l=1}^{n} \int_{C_D} d\omega_l \int_{\Gamma} \frac{dz_l}{2\pi i} \frac{b(\omega_l)}{b(z_l)} \right] W_-(\{\omega\}, \{z\})$$
$$\times \det\left[M(\omega_j, z_k) \right] \det\left[\frac{\rho(\omega_j, z_k) - \rho(\omega_j, \eta - z_k)}{2\,\text{sh}(2z_k - \eta)} \right]$$
(3.40)

with indices of the determinant $j, k = 1, \ldots, n$. C_D consists of the real line and an additional counterclockwisely closed contour around the pole $\omega = \eta/2 - \xi^-$ depending on the value of the boundary parameter ξ^-. The condition $-|\gamma|/2 < \text{Im}\,\xi^- < 0$ to include this additional contribution can easily be seen from figure 3.4, right panel.

Chapter 4

Separation of Variables

The hamiltonian (1.1) with full parameter range explicitly includes non-diagonal boundary contributions. Hence it exhibits a lower symmetry compared to the diagonal case and the spectrum can generally not be determined by the standard procedure of the algebraic Bethe ansatz. Although there have been various attempts to study the spectrum of this spin chain a satisfactory general scheme is still missing.

In this chapter we will first apply the functional Bethe ansatz or the separation of variables method elaborated by Sklyanin [84] to the transfer matrix corresponding to the spin chain with non-diagonal boundaries. Within this approach the eigenvalue problem is formulated using a suitably chosen representation of the underlying Yang-Baxter algebra on a space of symmetric functions. The main result is a TQ-equation similar to Baxter's approach but only valid for values of the spectral parameter on a certain grid which in return allows numerical solutions for small system sizes. Unfortunately to arrive at this result for unrestricted boundary parameters we need to restrict ourselves to the XXX model at some point in the calculation.

The algebraic results for the XXX spin chain are also valid for the so-called spin-boson model as it arises from the same R-matrix within QISM. The spin-boson model is a two site model coupling a spin-$\frac{1}{2}$ and a bosonic degree of freedom to each other and possibly to boundary fields. In section 4.3 the details arising from this situation are worked out resulting again in a TQ-equation valid on a grid.

4. SEPARATION OF VARIABLES

4.1 Functional Bethe Ansatz

The starting point of the method is the representation $U(\lambda)$ (2.35) of the left reflection algebra containing all the operator content. The entries of this monodromy matrix $U(\lambda)$ can be expressed in terms of the operators $A(\lambda)$, $B(\lambda)$, $C(\lambda)$ and $D(\lambda)$ of the periodic monodromy matrix $T(\lambda)$, i.e.

$$\begin{aligned}\mathcal{B}(\lambda) = &-\frac{\operatorname{sh}(\lambda - \frac{\eta}{2} + \xi^-)}{\operatorname{ch}(\lambda - \frac{\eta}{2})\operatorname{sh}\xi^-}\frac{\operatorname{sh}(2\lambda - \eta)}{\operatorname{sh}(2\lambda)}B(-\lambda - \frac{\eta}{2})A(\lambda - \frac{\eta}{2})\\&-\frac{\operatorname{sh}(\lambda + \frac{\eta}{2} - \xi^-)}{\operatorname{ch}(\lambda - \frac{\eta}{2})\operatorname{sh}\xi^-}\frac{\operatorname{sh}(2\lambda - \eta)}{\operatorname{sh}(2\lambda)}B(\lambda - \frac{\eta}{2})A(-\lambda - \frac{\eta}{2})\\&+\frac{\kappa^- e^{\theta^-}}{\operatorname{sh}\xi^-}\frac{\operatorname{sh}(2\lambda - \eta)}{\operatorname{ch}(\lambda - \frac{\eta}{2})}A(\lambda - \frac{\eta}{2})A(-\lambda - \frac{\eta}{2})\\&-\frac{\kappa^- e^{\theta^-}}{\operatorname{sh}\xi^-}\frac{\operatorname{sh}(2\lambda - \eta)}{\operatorname{ch}(\lambda - \frac{\eta}{2})}B(\lambda - \frac{\eta}{2})B(-\lambda - \frac{\eta}{2}).\end{aligned} \quad (4.1)$$

The presence of the term in the third line of (4.1) proportional to only A operators prevents the existence of an easy pseudo vacuum ($\mathcal{B}(\lambda)|0\rangle = 0$). But this term is no hindrance for Sklyanin's functional Bethe ansatz.

As mentioned above we will need to restrict ourselves to the rational case in order to have the boundary parameters unrestricted in the end. Then we can choose the right boundary matrix diagonal in favour of a twisted monodromy matrix $T(\lambda)$. Applying the rational limit the boundary matrix $K(\lambda, +)$ (2.30) yields the similarity transformation independent of the spectral parameter

$$K(\lambda, +) = MS\begin{pmatrix}\frac{\alpha^+ + \lambda}{\alpha^+} & 0 \\ 0 & \frac{\alpha^+ - \lambda}{\alpha^+}\end{pmatrix}(MS)^{-1} \quad (4.2)$$

with the diagonal matrix being a solution to the right reflection algebra and the 2×2 number matrices

$$M = \begin{pmatrix}e^{+\theta^+/2} & 0 \\ 0 & e^{-\theta^+/2}\end{pmatrix}, \quad S = \frac{1}{\sqrt{2\operatorname{ch}\beta^+}}\begin{pmatrix}e^{\beta^+/2} & -e^{-\beta^+/2} \\ e^{-\beta^+/2} & e^{\beta^+/2}\end{pmatrix} \quad (4.3)$$

being representations of the rational Yang-Baxter algebra. Thus in the transfer matrix $t(\lambda) = \operatorname{tr} K^{(+)}(\lambda) T(\lambda) K^{(-)}(\lambda) T^{-1}(-\lambda)$ we are free to consider a diagonal outer boundary matrix $K^{(+)}$ together with the c-number twist $(M^{(+)}S^{(+)})^{-1}T(\lambda)$ of $T(\lambda)$.[1]

4.1.1 Operator-Valued Zeros

The main goal of the functional Bethe ansatz is to treat the spectral problem of the transfer matrix in a representation space of symmetric functions manipulated by some shift opera-

---
[1] The $gl(2)$ symmetry of the rational model allows to remove the twist of the monodromy matrix in favour of a twisted boundary matrix $\widetilde{K}^{(-)} = (M^{(+)}S^{(+)})^{-1}K^{(-)}M^{(+)}S^{(+)}$, see Ref. [68].

4.1. Functional Bethe Ansatz

tors, which descent from operator-valued zeros of the $\mathscr{B}$-operator. Starting from $\mathscr{B}(\lambda)$ in the rational limit reading

$$\mathscr{B}(\lambda) = -\frac{2\lambda - ic}{\xi^-} \left[\frac{\lambda + \xi^- - \frac{ic}{2}}{2\lambda} B(-\lambda - \frac{ic}{2}) A(\lambda - \frac{ic}{2}) \right.$$
$$+ \frac{-\lambda + \xi^- - \frac{ic}{2}}{-2\lambda} B(\lambda - \frac{ic}{2}) A(-\lambda - \frac{ic}{2})$$
$$- \kappa^- e^{\theta^-} A(\lambda - \frac{ic}{2}) A(-\lambda - \frac{ic}{2})$$
$$\left. + \kappa^- e^{-\theta^-} B(\lambda - \frac{ic}{2}) B(-\lambda - \frac{ic}{2}) \right] \tag{4.4}$$

we observe the expression in the square brackets to be symmetric with respect to $\lambda \to -\lambda$ and having no pole at $\lambda = 0$. As the operators A, B, C and D are polynomials[2] in λ of degree L with the known asymptotics easily obtainable from their definition

$$A \sim \frac{\exp(\frac{\beta^+ - \theta^+}{2})}{\sqrt{2\operatorname{ch}\beta^+}} \lambda^L \quad , \quad B \sim \frac{\exp(\frac{\theta^+ - \beta^+}{2})}{\sqrt{2\operatorname{ch}\beta^+}} \lambda^L \;,$$
$$C \sim -\frac{\exp(-\frac{\beta^+ + \theta^+}{2})}{\sqrt{2\operatorname{ch}\beta^+}} \lambda^L \quad , \quad D \sim \frac{\exp(\frac{\beta^+ + \theta^+}{2})}{\sqrt{2\operatorname{ch}\beta^+}} \lambda^L \;, \tag{4.5}$$

we can factorize the square brackets in terms of (λ^2). The asymptotic prefactors arise from the twist $S^{-1}M^{-1}$ of the periodic monodromy matrix $T(\lambda)$.

Thus $\mathscr{B}(\lambda)$ is polynomial with a simple zero at $\lambda = ic/2$ and operator-valued coefficients assembling

$$\mathscr{B}(\lambda) = -\frac{2\lambda - ic}{(-1)^L \xi^-} \frac{1 - 2\kappa^- \operatorname{sh}(\theta^- - \theta^+ - \beta^+)}{2\operatorname{ch}\beta^+} \left[\prod_{\ell=1}^{L} (\lambda^2 - \hat{x}_\ell^2) \right]. \tag{4.6}$$

As $[\mathscr{B}(\lambda), \mathscr{B}(\mu)] = 0$, according to the reflection algebra, we can deduce $[\hat{x}_j^2, \hat{x}_k^2] = 0$ for all $j, k = 1 \ldots L$ with the spectrum shown in the next example.

Example. *For a spin-$\frac{1}{2}$ representation and $L = 1$ the explicit expression (4.6) with the inhomogeneity s_1 yields for the argument $\lambda = 0$ the form*

$$\hat{x}_1^2 = \begin{pmatrix} (s_1 + \frac{ic}{2})^2 & \\ & (s_1 - \frac{ic}{2})^2 \end{pmatrix} \tag{4.7}$$

of the operator-valued zero $\hat{x}_1^2$ in a diagonalized form. This fixes discrete sets $\Lambda_j \equiv \{s_j - ic/2, s_j + ic/2\}$ representing the spectra of the coordinates $\hat{x}_j$ except for a global sign and resembling $\mathbb{X}^L \equiv \Lambda_1 \times \ldots \times \Lambda_L$.

---
[2] From here the ε^L-dependence of all periodic chain operators is suppressed.

4. Separation of Variables

For the forthcoming relations, to work with the simple zeros $\hat{x}_j$ instead of $\hat{x}_j^2$, we refer to the following supplemental remark.

Remark. Let $\hat{x}_j$ be the operator-valued zeros satisfying $_{\lambda=\pm\hat{x}_j}|\mathscr{B}(\lambda) = 0$. Then all $\hat{x}_j^2$ can be simultaneously diagonalized such that for all $j,k = 1\ldots L$

$$[\hat{x}_j, \hat{x}_k] = 0 \quad , \quad \hat{x}_j = \begin{pmatrix} s_j + \frac{ic}{2} & \\ & s_j - \frac{ic}{2} \end{pmatrix}. \tag{4.8}$$

4.1.2 Conjugated Momenta

With the operators $\hat{x}_j$ at hand the next problem is to calculate the expression for the transfer matrix in the $\hat{x}$-representation. To do so let us first introduce the 'conjugated momenta' to the 'coordinates' $\hat{x}_j$.

Considering $\mathscr{A}(\lambda)$ and $\widetilde{\mathscr{D}}(\lambda)$ as polynomials and inserting the operator valued zeros of $\mathscr{B}(\lambda)$ by 'substitution from the left' yields the new operators

$$\begin{aligned}
_{\lambda=\hat{x}_j}|\mathscr{A}(\lambda) &= \sum_p \hat{x}_j^p \mathscr{A}_p \equiv X_j^- \\
_{\lambda=\hat{x}_j}|\widetilde{\mathscr{D}}(\lambda) &= \sum_p \hat{x}_j^p \widetilde{\mathscr{D}}_p \equiv X_j^+ .
\end{aligned} \tag{4.9}$$

Here $\mathscr{A}_p$ and $\widetilde{\mathscr{D}}_p$ denote operator-valued expansion coefficients. In the following theorem the commutation relations with the coordinates $\hat{x}_j$ are summarized.

Theorem 1. Let $\hat{x}_j$ be the operator-valued zeros of $\mathscr{B}(\lambda)$ and $X_j^\pm$ their conjugated momenta related by the reflection algebra. Then

$$X_j^\pm \hat{x}_k = (\hat{x}_k \pm ic\delta_{jk}) X_j^\pm . \tag{4.10}$$

Proof. Consider the commutation relation

$$\begin{aligned}
\mathscr{A}(\lambda)\mathscr{B}(\mu) =\,& \frac{\sh(\lambda+\mu-\eta)\sh(\lambda-\mu-\eta)}{\sh(\lambda+\mu)\sh(\lambda-\mu)} \mathscr{B}(\mu)\mathscr{A}(\lambda) \\
&+ \frac{\sh\eta\,\sh(2\mu-\eta)}{\sh(2\mu)\sh(\lambda+\mu)} \mathscr{B}(\lambda)\mathscr{A}(\mu) \\
&- \frac{\sh\eta}{\sh(2\mu)\sh(\lambda+\mu)} \mathscr{B}(\lambda)\widetilde{\mathscr{D}}(\mu)
\end{aligned} \tag{4.11}$$

of $\mathscr{A}(\lambda)$ and $\mathscr{B}(\mu)$ and multiply it by $\sh(\lambda+\mu)\sh(\lambda-\mu)$. Then take its rational limit and insert the coordinates $\hat{x}_j$ by 'substitution from the left'. The expression reduces to

$$(\hat{x}_j - \lambda)(\hat{x}_j + \lambda) X_j^- \mathscr{B}(\lambda) = (\hat{x} + \lambda - ic)(\hat{x} - \lambda - ic) \mathscr{B}(\lambda) X_j^- . \tag{4.12}$$

4.1. Functional Bethe Ansatz

Replacing $\mathscr{B}(\lambda)$ by its factorized form, cancelling the constant asymptotics and multiplying by the inverse $(\widehat{x}_j^2 - \lambda^2)^{-1}$ from the left the commutation relation

$$X_j^- \left[\prod_{\ell=1}^{L} (\lambda^2 - \widehat{x}_\ell^2) \right] = \left[\lambda^2 - (\widehat{x}_j - \mathrm{i}c)^2 \right] \left[\prod_{\substack{\ell=1 \\ \ell \neq j}}^{L} (\lambda^2 - \widehat{x}_\ell^2) \right] X_j^- \tag{4.13}$$

remains. Implying all expressions to be symmetric in $\widehat{x}_j$ to act on we arrive at the desired relation. Analogously the elementary commutation of X_j^+ with the coordinates arises from the commutation of $\mathscr{B}$ and $\widetilde{\mathscr{D}}$. □

The next natural step would be to establish the commutation relation between two $X^\pm$ operators. However, this cannot be done directly because per definition $X_j^\pm$ exceed the representation space.

4.1.3 Representation Space

Following Sklyanin's approach the square brackets in operator (4.6) can be expanded into $\lambda^{2L} - \widehat{b}_1 \lambda^{2(L-1)} \pm \ldots + \widehat{b}_L$ with commuting operators $\widehat{b}_j$ thus sharing a common system of eigenfunctions f_α,

$$\widehat{b}_j f_\alpha = b_j^\alpha f_\alpha \quad , \quad \alpha = 1 \ldots 2^L \tag{4.14}$$

where 2^L represents spin-$\frac{1}{2}$. To every point $\mathbf{b}^\alpha = (b_1^\alpha, \ldots, b_L^\alpha) \in \mathbb{B}^L \subset \mathbb{C}^L$ there corresponds only one eigenfunction f_α and the representation space W (e.g. for the XXX chain we have $W = (\mathbb{C}^2)^{\otimes L}$) is isomorphic to the space Fun $\mathbb{B}^L$.

Example. *A possible realization of the eigenfunctions f_α is*

$$(\widehat{b}_j f_\alpha)(\mathbf{b}^\beta) = b_j^\alpha f_\alpha(\mathbf{b}^\beta) \tag{4.15}$$

where $\widehat{b}_j$ act as multiplication operators. Let $\{\mathbf{t}_\alpha \in (\mathbb{C}^2)^{\otimes L} | (\mathbf{t}_\alpha)^\beta = f_\alpha(\mathbf{b}^\beta)\}$ be a basis of W then, with the constraint $f_\alpha(\mathbf{b}^\beta) = \delta_\alpha^\beta$, it is indeed orthonormal and complete.

Since $\widehat{b}_n$ are the symmetric polynomials of the roots $\widehat{x}_j^2$ we are led to consider the mapping

$$\theta : \mathbb{C}^L \to \mathbb{C}^L, \quad \mathbf{x} \mapsto \mathbf{b} \tag{4.16}$$

given by the formula $b_n(\mathbf{x}) = s_n(\mathbf{x})$. The $s_n(\mathbf{x})$ are the elementary symmetric polynomials of degree $n = 1 \ldots L$ of c-number variables

$$\begin{aligned} s_1(\mathbf{x}) &= x_1^2 + x_2^2 + \ldots + x_L^2 \\ &\vdots \\ s_L(\mathbf{x}) &= x_1^2 x_2^2 \ldots x_L^2. \end{aligned} \tag{4.17}$$

The diagram

4. SEPARATION OF VARIABLES

$$\mathbb{X}^L \xrightarrow{\theta} \mathbb{B}^L \xrightarrow{f} \{0,1\} \subset \mathbb{C}$$

with g on top and $f \circ \theta$ below.

of the combined mapping $f \circ \theta$ reveals the isomorphism between $\operatorname{Fun} \mathbb{B}^L \cong W$ and the space of symmetric functions $\operatorname{SymFun} \mathbb{X}^L$. The set $\{0,1\}$ is the range of f in the example above. Thus the operator roots $\hat{x}_j^2$ can be thought of as multiplication operators

$$\hat{x}_j^2 g(y_1, \ldots, y_L) = y_j^2 g(y_1, \ldots, y_L) \tag{4.18}$$

in an extended representation space $\operatorname{Fun} \mathbb{X}^L \cong \widetilde{W}$ which is a non-physical one. Recall all the results should only use the original space $\operatorname{SymFun} \mathbb{X}^L \cong W$, as the operators $\mathscr{A}$, $\mathscr{B}$, $\mathscr{C}$ and $\mathscr{D}$ map $\operatorname{SymFun} \mathbb{X}^L \to \operatorname{SymFun} \mathbb{X}^L$ or more sloppy $W \to W$. So for plainness we will use in the following the terms W and $\widetilde{W}$ for the representation spaces instead.

For the action of $X_j^\pm$ on a function $s \in \operatorname{SymFun} \mathbb{X}^L$ we need first to extend the operators from W to $\widetilde{W}$ by the constant function

$$\omega(\mathbf{x}) = 1 \text{ for all } \mathbf{x} \in \mathbb{X}^L. \tag{4.19}$$

Obviously ω is symmetric and thus belongs to the representation space W. Now define the action of $X_j^\pm$ on ω by

$$(X_j^\pm \omega)(\mathbf{x}) \equiv \Delta_j^\pm(\mathbf{x}). \tag{4.20}$$

Then the functions $\Delta_j^\pm(\mathbf{x})$ uniquely determine the action of $X_j^\pm$ on any vector s which is identified due to the isomorphism with some symmetric function $s(x_1, \ldots, x_L) = (\hat{s}\omega)(\mathbf{x})$ created from the cyclic vector ω by the operator $\hat{s} = s(\hat{x}_1, \ldots, \hat{x}_L)$. Thus

$$(X_j^\pm s)(\mathbf{x}) = (X_j^\pm \hat{s}\omega)(\mathbf{x}) = s(E_j^\pm \mathbf{x})(X_j^\pm \omega)(\mathbf{x}) = s(E_j^\pm \mathbf{x})\Delta_j^\pm(\mathbf{x}). \tag{4.21}$$

Here we introduced the shift operators

$$E_j^\pm : \mathbb{C}^L \to \mathbb{C}^L : (x_1, \ldots, x_j, \ldots, x_L) \mapsto (x_1, \ldots, x_j \pm ic, \ldots, x_L) \tag{4.22}$$

acting on some L-tuple of c-numbers. In the extended representation space $\widetilde{W}$ of not necessarily symmetric functions the action of $X_j^\pm$ then reads

$$X_j^\pm = \Delta_j^\pm E_j^\pm \tag{4.23}$$

with $\Delta_j^\pm = \Delta_j^\pm(\mathbf{x})$. By the operator relation (4.23) we can now calculate the commutations of the momenta.

Theorem 2. *Let $X_j^\pm$ be the conjugated momenta related to the coordinates $\hat{x}_j$ by the reflection algebra. Then*

$$\begin{aligned} \left[X_j^\pm, X_k^\pm\right] &= 0, \quad \text{for all } j,k = 1 \ldots L \\ \left[X_j^+, X_k^-\right] &= 0, \quad \text{for all } j,k = 1 \ldots L \text{ but } j \neq k. \end{aligned} \tag{4.24}$$

4.1. Functional Bethe Ansatz

Proof. Let us start with X^- where the first assertion is obvious for $j = k$. Then it is enough to consider the cases $j = 1$, $k = 2$. Taking the rational limit of

$$[\mathscr{A}(\lambda), \mathscr{A}(\mu)] = \frac{\operatorname{sh}\eta}{\operatorname{sh}(\lambda + \mu)} [\mathscr{B}(\mu)\mathscr{C}(\lambda) - \mathscr{B}(\lambda)\mathscr{C}(\mu)] \quad (4.25)$$

and inserting $\lambda = \hat{x}_1$ and $\mu = \hat{x}_2$ by 'substitution from the left' the right hand side turns into zero and for the left hand side we get

$$\left.{}_{\lambda=\hat{x}_1, \mu=\hat{x}_2}\right| \mathscr{A}(\lambda)\mathscr{A}(\mu) = \sum_{m,n} \hat{x}_1^m \hat{x}_2^n \mathscr{A}_m \mathscr{A}_n = \sum_{m,n} \hat{x}_2^n \hat{x}_1^m \mathscr{A}_m \mathscr{A}_n \\
= \sum_n \hat{x}_2^n X_1^- \mathscr{A}_n = X_1^- \sum_n \hat{x}_2^n \mathscr{A}_n = X_1^- X_2^- . \quad (4.26)$$

In the same way starting from $\mathscr{A}(\mu).\mathscr{A}(\lambda)$ one obtains $X_2^- X_1^-$ and the assertion is proven. The commutation of X^+'s and mixed commutators excluding the cases $j = k$ can be treated analogously by considering

$$\left[\widetilde{\mathscr{D}}(\lambda), \widetilde{\mathscr{D}}(\mu)\right] = -\operatorname{sh}(2\lambda + \eta)\operatorname{sh}(2\mu + \eta)\left[\mathscr{A}(\lambda), \mathscr{A}(\mu)\right] \\
\left[\widetilde{\mathscr{D}}(\lambda), \mathscr{A}(\mu)\right] = \frac{\operatorname{sh}(\lambda+\mu)\operatorname{sh}(2\lambda+\eta)}{\operatorname{sh}(\lambda-\mu)}\left[\mathscr{A}(\lambda), \mathscr{A}(\mu)\right] \quad (4.27)$$

in the rational limit. $\square$

The remaining commutation relation involving the quantum determinant Δ_q is summarized in the following theorem.

Theorem 3. *Let $\hat{x}_j$ and $X_j^\pm$ be the coordinates and conjugated momenta related by the reflection algebra and $\Delta_q(\lambda)$ is the quantum determinant. Then*

$$X_j^\pm X_j^\mp = \Delta_q(\hat{x}_j \pm \tfrac{\mathrm{i}c}{2}) \text{ for all } j, k = 1\ldots L. \quad (4.28)$$

Proof. Substituting the operator-valued zeros $\hat{x}_j$ into the suggestive form (2.41) of the quantum determinant one obtains

$$\Delta_q(\hat{x}_j - \tfrac{\mathrm{i}c}{2}) = \sum_{m,n} \hat{x}_j^m (\hat{x}_j - \mathrm{i}c)^n \mathscr{A}_m \widetilde{\mathscr{D}}_n = \sum_{m,n} (\hat{x}_j - \mathrm{i}c)^n \left(\hat{x}_j^m \mathscr{A}_m\right) \widetilde{\mathscr{D}}_n \\
= \sum_n (\hat{x}_j - \mathrm{i}c)^n X_j^- \widetilde{\mathscr{D}}_n = X_j^- \sum_n \hat{x}_j^n \widetilde{\mathscr{D}}_n \quad (4.29) \\
= X_j^- X_j^+$$

and analogously $\Delta_q(\hat{x}_j + \tfrac{\mathrm{i}c}{2}) = X_j^+ X_j^-$ exerting the reflection algebra. $\square$

Remark. *The remaining zero* $\mathrm{i}c/2$ *of* $\mathscr{B}(\lambda)$ *is an exception and renders the operators* $\mathscr{A}(\mathrm{i}c/2) = (d_q T)(-\mathrm{i}c/2)$ *and* $\widetilde{\mathscr{D}}(\mathrm{i}c/2) = 0$ *to be constant yielding* $\Delta_q(\mathrm{i}c) = 0$.

4.1.4 Representation of $\Delta^{\pm}$

Applying $X_j^{\pm} X_j^{\mp} = \Delta_j^{\pm} E_j^{\pm} \Delta_j^{\mp} E_j^{\mp}$ to an arbitrary function $g \in \widetilde{W}$ induces the sequence

$$\begin{aligned}(X_j^{\pm} X_j^{\mp} g)(\mathbf{x}) &= (\Delta_j^{\pm} E_j^{\pm} \Delta_j^{\mp} E_j^{\mp} g)(\mathbf{x}) = \Delta_j^{\pm}(\mathbf{x})(E_j^{\pm} \Delta_j^{\mp} E_j^{\mp} g)(\mathbf{x}) \\ &= \Delta_j^{\pm}(\mathbf{x})(\Delta_j^{\mp} E_j^{\mp} g)(E_j^{\pm} \mathbf{x}) = \Delta_j^{\pm}(\mathbf{x}) \Delta_j^{\mp}(E_j^{\pm} \mathbf{x})(E_j^{\mp} g)(E_j^{\pm} \mathbf{x}) \\ &= \Delta_j^{\pm}(\mathbf{x}) \Delta_j^{\mp}(E_j^{\pm} \mathbf{x}) g(\mathbf{x}) \\ &\overset{!}{=} \Delta_q(x_j \pm \tfrac{ic}{2}) g(\mathbf{x}) \end{aligned} \quad (4.30)$$

relating the representations $\Delta_j^{\pm}$ to the quantum determinant Δ_q. In the case of a finite dimensional representation of the generators $\{\widehat{x}_j, X_j^{\pm}\}_{j=1}^{L}$ such that the spectrum $\mathbb{X}^L$ shows no multiple points the problem of constructing such a representation is equivalent to that of determining the functions $\{\Delta_j^{\pm}\}_{j=1}^{L}$ on $\mathbb{X}^L$ satisfying

$$\begin{aligned} \Delta_m^{\pm}(\mathbf{x}) \Delta_n^{\pm}(E_m^{\pm} \mathbf{x}) &= \Delta_n^{\pm}(\mathbf{x}) \Delta_m^{\pm}(E_n^{\pm} \mathbf{x}) & \text{for all } n,m \\ \Delta_m^{+}(\mathbf{x}) \Delta_n^{-}(E_m^{+} \mathbf{x}) &= \Delta_n^{-}(\mathbf{x}) \Delta_m^{+}(E_n^{-} \mathbf{x}) & \text{for all } n,m \text{ but } n \neq m \\ \Delta_q(\widehat{x}_j \pm \tfrac{ic}{2}) &= \Delta_j^{\pm}(\mathbf{x}) \Delta_j^{\mp}(E_j^{\pm} \mathbf{x}) & \text{for all } j \end{aligned} \quad (4.31)$$

arising from theorems 2 and 3. The above relations are not defined when the shifts $E_j^{\pm}$ move the point $\mathbf{x}$ out of $\mathbb{X}^L = \Lambda_1 \times \ldots \times \Lambda_L$. This means $\{\Delta_j^{\pm}\}_{j=1}^{L}$ have to vanish on the boundary

$$\partial \mathbb{X}_j^{\pm} \equiv \{\mathbf{x} \in \mathbb{X}^L | E_j^{\pm} \mathbf{x} \in \mathbb{C}^L \setminus \mathbb{X}^L\} \quad (4.32)$$

of the set $\mathbb{X}^L$. For the open XXX chain with $\Lambda_j = \{s_j - ic/2, s_j + ic/2\}$ this is clear from the explicit factorization of the quantum determinant.

Example. *The vanishing of $\Delta_j^{\pm}(\mathbf{x})$ on the boundary $\partial \mathbb{X}_j^{\pm}$ can be directly seen from the explicit factorization of $\Delta_q(\lambda) = \Delta^{-}(\lambda + \eta/2)\Delta^{+}(\lambda - \eta/2)$ into*

$$\begin{aligned} \Delta^{-}(\lambda) &= \frac{\lambda - \tfrac{ic}{2} + \alpha^{-}}{(-1)^L \alpha^{-}} \Big[\prod_{l=1}^{L} (\lambda - s_l + \tfrac{ic}{2})(\lambda + s_l + \tfrac{ic}{2}) \Big] \\ \Delta^{+}(\lambda) &= -(2\lambda - ic)\varepsilon \frac{\lambda + \tfrac{ic}{2} - \alpha^{-}}{(-1)^L \alpha^{-}} \Big[\prod_{l=1}^{L} (\lambda - s_l - \tfrac{ic}{2})(\lambda + s_l - \tfrac{ic}{2}) \Big] \end{aligned} \quad (4.33)$$

considered in the rational limit indicated by $\varepsilon \to 0$.

4.1.5 Spectral Analysis

Now let us return to the original problem, the spectral analysis of the scaled transfer matrix (2.37), $\tau(\lambda) = \text{tr} K(\lambda + \eta/2, +) U(\lambda)/2$, in the rational limit

$$\tau(\lambda) = \frac{(\lambda + \tfrac{ic}{2})(\lambda + \xi^{+} - \tfrac{ic}{2})}{2\lambda \xi^{+}} \mathscr{A}(\lambda) - \frac{1}{\varepsilon} \frac{\lambda - \xi^{+} + \tfrac{ic}{2}}{4\lambda \xi^{+}} \widetilde{\mathscr{D}}(\lambda) + \frac{(\lambda + \tfrac{ic}{2})\kappa^{+}}{\xi^{+}} \Big[e^{\theta^{+}} \mathscr{C}(\lambda) + e^{-\theta^{+}} \mathscr{B}(\lambda) \Big] \quad (4.34)$$

4.1. Functional Bethe Ansatz

and mind the scaling factor $\varepsilon \to 0$. To plug in the zeros $\hat{x}_j$ by 'substitution from the left' we have to get rid of $\mathscr{C}(\lambda)$ by diagonalizing $K^{(+)}$. Thus only the first line remains. The diagonalization does not change the quantum determinant $\Delta_q(\lambda)$ and the eigenvalue problem $\tau(\lambda)\varphi = \Lambda(\lambda)\varphi$ can be solved by 'substitution from the left' reading

$$\left.|\tau(\lambda)\right|_{\lambda=\hat{x}_j} = \frac{(\hat{x}_j + \frac{ic}{2})(\hat{x}_j + \alpha^+ - \frac{ic}{2})}{2\hat{x}_j \alpha^+} X_j^- - \frac{1}{\varepsilon} \frac{\hat{x}_j - \alpha^+ + \frac{ic}{2}}{4\hat{x}_j \alpha^+} X_j^+ \,. \tag{4.35}$$

With this representation at hand one observes 'separation of variables' suggesting the product ansatz

$$\varphi = \Big[\prod_{\ell=1}^{L} Q(x_\ell)\Big] \tag{4.36}$$

for the eigenfunction $\varphi \in \operatorname{SymFun} \mathbb{X}^L \cong W$ symmetric in its arguments x_ℓ. To explicitly apply the operator-valued expression (4.35) we need to clarify its behaviour on generally symmetric functions via (4.18) and (4.23).

Lemma. *The action of the combined expression $\hat{x}_j X_j^\pm$ by 'substitution from the left' onto a symmetric function $s = s(x_1, \ldots, x_L)$ is given by*

$$\hat{x}_j X_j^\pm s(\mathbf{x}) = (\hat{x}_j X_j^\pm s)(\mathbf{x}) = x_j (X_j^\pm s)(\mathbf{x}) = x_j \Delta_j^\pm(\mathbf{x}) s(E_j^\pm \mathbf{x}) \,. \tag{4.37}$$

Then applying (4.35) to φ only the jth argument is affected such that the problem separates and

$$\begin{aligned}\Lambda(x_j)Q(x_j) =& \frac{(x_j + \frac{ic}{2})(x_j + \alpha^+ - \frac{ic}{2})}{2x_j \alpha^+} \Delta^-(x_j) Q(x_j - ic) \\&- \frac{1}{\varepsilon} \frac{x_j - \alpha^+ + \frac{ic}{2}}{4 x_j \alpha^+} \Delta^+(x_j) Q(x_j + ic)\end{aligned} \tag{4.38}$$

holds. Here we used (4.37) with the allowed arguments $x_j \in \Lambda_j = \{s_j - ic/2, s_j + ic/2\}$ on the grid entering $\Delta_j^\pm(\mathbf{x}) = \Delta^\pm(x_j)$.

The eigenvalue problem as formulated in (4.38) reduces to a system of homogeneous linear equations due to the fact that $\Delta^\pm(x_j^\pm) = 0$ at the points $x_j^\pm = s_j \pm ic/2$:

$$\begin{aligned}\Lambda(x_j^+)Q(x_j^+) =& \frac{(x_j^+ + \frac{ic}{2})(x_j^+ + \alpha^+ - \frac{ic}{2})}{2x_j^+ \alpha^+} \Delta^-(x_j^+)Q(x_j^-) \\\Lambda(x_j^-)Q(x_j^-) =& -\frac{1}{\varepsilon} \frac{x_j^- - \alpha^+ + \frac{ic}{2}}{4 x_j^- \alpha^+} \Delta^+(x_j^-)Q(x_j^+) \,.\end{aligned} \tag{4.39}$$

For pairwise different inhomogeneities, $s_j \neq s_k$ for $j \neq k$, these linear equations allow for a non-trivial solution provided that the following functional equations for the eigenvalues Λ

are satisfied[3]

$$\Lambda(s_j + \tfrac{ic}{2})\Lambda(s_j - \tfrac{ic}{2}) = -\frac{s_j + ic}{2\varepsilon} \frac{s_j - \alpha^+}{(2s_j - ic)\alpha^+} \frac{s_j + \alpha^+}{(2s_j + ic)\alpha^+} \Delta_q(s_j), \quad j = 1\ldots L. \quad (4.40)$$

Using the known asymptotic form (2.86) of the even polynomial $\Lambda(\lambda) = \Lambda(-\lambda)$ we are led to the ansatz

$$\Lambda(\lambda) = \frac{(-1)^L}{\alpha^+ \alpha^-} \frac{\operatorname{sh}\beta^+ \operatorname{sh}\beta^- + \operatorname{ch}(\theta^+ - \theta^-)}{\operatorname{ch}\beta^+ \operatorname{ch}\beta^-} \lambda^{2L+2} + a_{2L}\lambda^{2L} + a_{2L-2}\lambda^{2L-2} + \ldots + a_0. \quad (4.41)$$

The $(L+1)$ unknown coefficients a_j are determined by eqs. (4.40) and the constraint $\Lambda(ic/2) = (d_q T)(-ic/2)$ with the quantum determinant $(d_q T)(\lambda)$ of the periodic chain. This immediately follows from the property $t(0) = 1$ of the unshifted and unscaled transfer matrix (2.33). Thus the solution of the spectral problem amounts to finding the common roots $\{a_{2j}^{(\nu)}\}_{j=0}^L$, $\nu = 1\ldots 2^L$, of these polynomial equations. This task is of the same complexity as finding the eigenvalues of the spin chain Hamiltonian directly and therefore this approach is limited to small system sizes where we have checked numerically that it does indeed yield the complete spectrum.

To compute the eigenvalue of the transfer matrix or the spin chain Hamiltonian in the thermodynamic limit $L \to \infty$ the functional equations introduced above need to be analyzed beyond the set $\mathbb{X}^L$ using explicitly the analytic properties of the functions therein.

Treating the s_j in (4.40) as a continuous variable and applying standard Fourier techniques explained further in appendix F one can compute $\ln \Lambda(\lambda)$ and thereby the corresponding eigenvalue of the spin chain Hamiltonian (1.1). For $|\alpha^\pm| > c/2$ one obtains

$$\begin{aligned} ic\frac{\partial \ln \Lambda}{\partial \lambda}(\tfrac{ic}{2}) = &\psi(\tfrac{|\alpha^+|}{2c}) - \psi(\tfrac{|\alpha^+|}{2c} + \tfrac{1}{2}) + \frac{c}{|\alpha^+|} + \psi(\tfrac{|\alpha^-|}{2c}) - \psi(\tfrac{|\alpha^-|}{2c} + \tfrac{1}{2}) + \frac{c}{|\alpha^-|} \\ &+ \pi - 2\ln 2 - 1 + (2 - 4\ln 2)L \end{aligned} \quad (4.42)$$

which is for imaginary $\alpha^\pm$ the known energy eigenvalue of the *XXX* spin chain with diagonal boundary fields [33] (ψ is the digamma function). However, the non-diagonal contributions and corrections of the order $1/L$ are not included. This is a consequence of neglecting the corrections to (4.40) away from the points s_j. Including these unknown corrections in the equations we obtain an equation being reminiscent of the first level of the fusion equations (2.85). This observation will be further pursued in chapter 6.

4.2 TQ-Equation

The analysis above leading to (4.40) has been based on the singular points of (4.38) at the boundaries $\partial \mathbb{X}^L$, i.e. points where one of the coefficients $\Delta^\pm$ vanishes. Instead we go back

---

[3] If n of the inhomogeneities coincide the $(n-1)$ derivatives of this equation at this value of s_j have to be taken into account in addition.

4.2. TQ-Equation

one step and consider now (4.38) for general arguments $x_j \to \lambda$. Formally, this is a second order difference equation reminiscent of Baxter's TQ-equation [9]. Away from the singular points there exist two independent solutions to (4.38) and one needs some information on the properties of the unknown functions $Q(\lambda)$ in this formulation of the spectral problem which has to be solved for polynomial eigenvalues $\Lambda(\lambda)$ of the transfer matrix.

In cases where a pseudo vacuum exists and the algebraic Bethe ansatz is applicable to solve the problem the Q-functions are known to be symmetric polynomials $Q(\lambda) = [\prod_{\ell=1}^M (\lambda - v_\ell)(\lambda + v_\ell)]$ with roots v_ℓ satisfying Bethe ansatz equations. Note that in these cases the constant function $\omega = 1$ introduced in the construction of the representation of the Yang-Baxter algebra on the space $\widetilde{W}$ can be identified with the pseudo vacuum $|0\rangle$.

In general, the TQ equation can be rewritten as a recursion relation

$$Q(\lambda + ic) = a(\lambda)Q(\lambda) + b(\lambda)Q(\lambda - ic) \tag{4.43}$$

for the function $Q(\lambda)$ or equivalently, with the auxiliary function $P(\lambda + ic) \equiv Q(\lambda)$,

$$\begin{pmatrix} Q(\lambda + ic) \\ P(\lambda + ic) \end{pmatrix} = \begin{pmatrix} a(\lambda) & b(\lambda) \\ 1 & 0 \end{pmatrix} \begin{pmatrix} Q(\lambda) \\ P(\lambda) \end{pmatrix}. \tag{4.44}$$

The coefficients $a(\lambda)$ and $b(\lambda)$ are obtained from the TQ-equation (4.38) and show constant asymptotics for large values of their arguments

$$a(\lambda) = -\frac{\Lambda(\lambda)}{\Delta^+(\lambda)} \frac{4\varepsilon\lambda\alpha^+}{\lambda - \alpha^+ + \frac{ic}{2}} \sim 2\frac{\operatorname{sh}\beta^+ \operatorname{sh}\beta^- + \operatorname{ch}(\theta^+ - \theta^-)}{\operatorname{ch}\beta^+ \operatorname{ch}\beta^-},$$

$$b(\lambda) = \frac{\Delta^-(\lambda)}{\Delta^+(\lambda)} \frac{2(\lambda + \frac{ic}{2})(\lambda + \alpha^+ - \frac{ic}{2})\varepsilon}{\lambda - \alpha^+ + \frac{ic}{2}} \sim -1. \tag{4.45}$$

This allows to solve the recursion relations in the asymptotic regime $|\lambda| \gg 1$ yielding

$$Q(\lambda + nic) = \frac{\lambda_1^n - \lambda_2^n}{\lambda_1 - \lambda_2} Q(\lambda + ic) - \lambda_1 \lambda_2 \frac{\lambda_1^{n-1} - \lambda_2^{n-1}}{\lambda_1 - \lambda_2} Q(\lambda). \tag{4.46}$$

Here n is an integer and $\lambda_{1,2} = e^{\pm\phi}$ are the eigenvalues of the asymptotical matrix of coefficients in (4.44) with ϕ defined by

$$\operatorname{ch}\phi = \frac{\operatorname{sh}\beta^+ \operatorname{sh}\beta^- + \operatorname{ch}(\theta^+ - \theta^-)}{\operatorname{ch}\beta^+ \operatorname{ch}\beta^-}. \tag{4.47}$$

Ordering the eigenvalues as $|\lambda_1| > |\lambda_2|$ we obtain for fixed λ and large n the leading term $Q(\lambda + nic) \sim \lambda_1^n$ suggesting the following ansatz for the asymptotic form

$$Q(\lambda) \sim \exp\left(\frac{\lambda\phi}{ic}\right) \times \ldots \times (\text{polynomial in } \lambda). \tag{4.48}$$

Here the polynomial form of the sub-leading part assures that the eigenvalue $\Lambda(\lambda)$ of the transfer matrix remains polynomial. Note that since $\Lambda(\lambda)$ is an even function of its argument

4. Separation of Variables

there exists always a second solution $Q(-\lambda)$ to the TQ-equation which decays exponentially for $\lambda \to \infty$.

Only in two cases, namely $\phi = 0$ and $i\pi$ or, equivalently,

$$\text{ch}(\theta^+ - \theta^-) = \pm \text{ch}(\beta^+ \mp \beta^-), \tag{4.49}$$

the exponential factor disappears and the TQ equation can be solved by an even polynomial: in the first case (4.48) implies that $Q(\lambda) = \prod_{\ell=1}^{M^{(+)}} (\lambda - v_\ell)(\lambda + v_\ell)$. For $\phi = i\pi$, the exponential factors can be removed by the transformation $Q(\lambda) = \exp(i\lambda\pi/ic)Q'(\lambda)$ resulting in a TQ equation for Q':

$$\Lambda(x_j)Q'(x_j) = -\frac{(x_j + \frac{ic}{2})(x_j + \alpha^+ - \frac{ic}{2})}{2x_j\alpha^+}\Delta^-(x_j)Q'(x_j - ic)$$
$$+ \frac{1}{\varepsilon}\frac{x_j - \alpha^+ + \frac{ic}{2}}{4x_j\alpha^+}\Delta^+(x_j)Q'(x_j + ic). \tag{4.50}$$

Again, it follows from the asymptotic analysis that this equation allows for a polynomial solution $Q'(\lambda) = \prod_{\ell=1}^{M^{(-)}} (\lambda - v_\ell)(\lambda + v_\ell)$ whose existence has been verified by numerical analysis for small systems.

In both cases the spectrum is determined by the roots of these polynomials. To guarantee analyticity of the transfer matrix eigenvalues $\Lambda(\lambda)$ the v_j, $j = 1\ldots M^{(\pm)}$, have to satisfy the Bethe ansatz equations

$$\frac{v_j + \alpha^- - \frac{ic}{2}}{v_j - \alpha^- + \frac{ic}{2}}\frac{v_j + \alpha^+ - \frac{ic}{2}}{v_j - \alpha^+ + \frac{ic}{2}}\left[\prod_{\ell=1}^{L}\frac{v_j - s_\ell + \frac{ic}{2}}{v_j - s_\ell - \frac{ic}{2}}\frac{v_j + s_\ell + \frac{ic}{2}}{v_j + s_\ell - \frac{ic}{2}}\right]$$
$$= \left[\prod_{\substack{k=1 \\ k \ne j}}^{M^{(\pm)}}\frac{v_j - v_k + ic}{v_j - v_k - ic}\frac{v_j + v_k + ic}{v_j + v_k - ic}\right]. \tag{4.51}$$

Note that (4.49) is equivalent to the constraint that the boundary matrices $K^{(\pm)}$ can be simultaneously diagonalized or brought to triangular form. In this case (4.51) can be obtained by means of the algebraic Bethe ansatz [68] or in the rational limit from the TQ-equation approach to the open XXZ chain [99]. In this trigonometric case the complete set of eigenvalues is obtained from two sets of Bethe equations which both reduce to (4.51) in the rational limit. This is due to the invariance of the model under the change of parameters $\alpha \to -\alpha$ and $\beta \to i\pi - \beta$ which maps $\phi = 0$ to $\phi = i\pi$ or vice versa, see (2.29). As another difference to the situation in the XXZ model the number of Bethe roots, and hence the degree of the corresponding Q-function, is not restricted by the constraint on the boundary fields and we have to consider solutions of the TQ-equations 'beyond the equator', $M^{(\pm)} > L/2$. We suppose that this feature of the XXX case is a consequence of the rational limit.

In contrast to earlier approaches to the XXZ chain by Murgan, Nepomechie et al. [71, 73] resulting in numerous TQ-equations for the general boundary parameters we find one single

TQ-equation (4.38) to determine the spectrum of the XXX chain. This supports the findings of Yang et al. [99] who also obtain a single TQ-equation for the XXZ model assuming the existence of a certain limit in the auxiliary space. Nevertheless it is interesting to study the differences of their approach to the functional Bethe ansatz and in particular the explicit form of the TQ-equation and the Q-function appearing in both methods. For this reason we will look at the XXZ model with non-diagonal *twisted* boundary conditions in chapter 5 because for this model both approaches are applicable.

4.3 Spin Boson Model

As a consequence of the above results it is possible to apply the functional Bethe ansatz to the spin boson model. This system consists of only two sites – one equipped with a spin-$\frac{1}{2}$ and the other with a bosonic degree of freedom. Both sites are coupled to each other with XXX-type interaction and are exposed to (non-diagonal) boundary fields. This setup is a toy model for a two-level atom coupled to a single mode of the quantized electromagnetic field in an ideal cavity. The model was first introduced by Jaynes, Tavis and Cummings [45,90,91] and is exactly diagonalizable for weak or resonant interactions by application of the *rotating wave approximation* (RWA) [41,45,90,91]. This approximation assumes that only coherent oscillations of the population of the atomic energy levels generate the relevant dynamics. For this reason only operators $a^\dagger S^-$ or aS^+ describing photon emission accompanied by atomic excitation or vice versa respectively are taken into account in the coupling to the electric field. These terms are called *rotating*. On the other hand the so-called *counter-rotating* terms $a^\dagger S^+$ and aS^- induce only short time dynamics and can be neglected at resonant cavity frequency in the RWA. Then the result is a conserved operator $S^z + a^\dagger a$ and a reduction of the originally infinite rank hamiltonian to a finite dimensional subspace is possible.

The spectral problem remains unsolved in cases where the RWA is not applicable e.g. due to effectively large coupling of the spin-boson to the electric field in superconducting circuits [94] or Rashba and Dresselhaus spin-orbit interactions in semiconductors [27]. In this case interesting hamiltonians contain rotating *and* counter-rotating terms simultaneously. Such hamiltonians were derived within QISM by Amico, Frahm et al. in [5] where the presence of non-diagonal boundary fields was found to be inevitable.

4.3.1 Construction of the Transfer Matrix

The hamiltonians including rotating and counter-rotating terms were derived within QISM with the bosonic Lax operator [5, 14]

$$\mathscr{L}^b(\lambda) = \begin{pmatrix} \lambda + (\frac{1}{2} + \delta_b)\mathrm{i}c - \mathrm{i}c\, a^\dagger a & \beta a^\dagger \\ \gamma a & -\frac{\beta\gamma}{\mathrm{i}c} \end{pmatrix} \qquad (4.52)$$

4. Separation of Variables

where δ_b is a inhomogeneity and $a^\dagger$ and a are the standard bosonic creation and annihilation operators. The scalar parameters β and γ remain unspecified. The representation of the Yang-Baxter algebra used for the spin degree of freedom was introduced in (2.11). To use notation being reminiscent of the notation of [5] we scale it and shift the spectral parameter

$$\mathscr{L}^s(\lambda) = \begin{pmatrix} \lambda + s\,\mathrm{ic} + \mathrm{ic}\,S^z & \mathrm{ic}\,S^- \\ \mathrm{ic}\,S^+ & \lambda + s\,\mathrm{ic} - \mathrm{ic}\,S^z \end{pmatrix}. \tag{4.53}$$

The parameter s controls the introduced shift and with $s = 1$ we recover the Lax operator from [5]. Further we want to introduce the inhomogeneity δ_s on the spin site via $s = \frac{1}{2} + \delta_s$. The quantum determinants of these Lax operators are $d_q(\mathscr{L}^b)(\lambda) = -\frac{\beta\gamma}{\mathrm{ic}}(\lambda + (\delta_b + 1)\mathrm{ic})$ and $d_q(\mathscr{L}^s)(\lambda) = (\lambda + s\,\mathrm{ic} - \mathrm{ic})(\lambda + s\,\mathrm{ic} + \mathrm{ic})$. The corresponding periodic monodromy matrix $T(\lambda)$ obeying the Yang-Baxter algebra (2.1) is

$$T(\lambda) = \mathscr{L}^b(\lambda) \cdot \mathscr{L}^s(\lambda) = \begin{pmatrix} A(\lambda) & B(\lambda) \\ C(\lambda) & D(\lambda) \end{pmatrix} \tag{4.54}$$

and enters the scaled open boundary transfer matrix as given in (2.37).

4.3.2 Functional Bethe Ansatz

The separation of variables method starts with a factorization of the $\mathscr{B}$ operator of the monodromy matrix $U(\lambda)$ using its operator valued zeros. Of course the algebraic background does not differ from the XXX spin chain and hence we only need to identify the correct 'coordinates' $\hat{x}$ and their conjugated momenta. Recalling the expression of $\mathscr{B}$ in terms of periodic chain operators from (4.4)

$$\mathscr{B}(\lambda) = -\frac{2\lambda - \mathrm{ic}}{\xi^-}\left[\frac{\lambda + \xi^- - \frac{\mathrm{ic}}{2}}{2\lambda} B(-\lambda - \tfrac{\mathrm{ic}}{2})A(\lambda - \tfrac{\mathrm{ic}}{2})\right.$$
$$+ \frac{-\lambda + \xi^- - \frac{\mathrm{ic}}{2}}{-2\lambda} B(\lambda - \tfrac{\mathrm{ic}}{2})A(-\lambda - \tfrac{\mathrm{ic}}{2})$$
$$\left. - \kappa^- e^{\theta^-} A(\lambda - \tfrac{\mathrm{ic}}{2})A(-\lambda - \tfrac{\mathrm{ic}}{2}) + \kappa^- e^{-\theta^-} B(\lambda - \tfrac{\mathrm{ic}}{2})B(-\lambda - \tfrac{\mathrm{ic}}{2})\right]$$

the expression in the bracket is symmetric with respect to the sign of λ and will be referred to as $\mathscr{B}_{\mathrm{symm}}$. The expansion of $\mathscr{B}$ in the operator valued zeros $\hat{x}$ and the c-number valued zero $\frac{\mathrm{ic}}{2}$ for this model reads

$$\mathscr{B}(\lambda) = \frac{2\lambda - \mathrm{ic}}{\xi^-} \frac{1 - 2\kappa^- \,\mathrm{sh}(\theta^- - \theta^+ - \beta^+)}{2\,\mathrm{ch}\,\beta^+}(\lambda^2 - \hat{x}_s)(\lambda^2 - \hat{x}_b). \tag{4.55}$$

The operator valued zeros are then determined by the expansion of

$$\mathscr{B}_{\mathrm{symm}}(\lambda) = B_4 \lambda^4 + B_2 \lambda^2 + B_0 \tag{4.56}$$

4.3. Spin Boson Model

with B_i being coefficients involving spin and bosonic operators. Using symmetric polynomials $b_1 = -B_2/B_4$ and $b_2 = B_0/B_4$ we obtain

$$\hat{x}_b^2 + \hat{x}_s^2 = b_1 \quad , \quad \hat{x}_b^2 \cdot \hat{x}_s^2 = b_2 \,. \tag{4.57}$$

The explicit coefficients b_1 and b_2 were computed using FORM [93].

For the detailed analysis we will first focus on b_1. The bosonic Lax operator has entries proportional to a single creation operator $a^\dagger$. As $a^\dagger$ does not have any right-eigenstates but a does we will consider $b_1^\dagger$ for convenience. Choosing the standard-basis $|\uparrow\rangle \equiv \begin{pmatrix} 1 \\ 0 \end{pmatrix}$ and $|\downarrow\rangle \equiv \begin{pmatrix} 0 \\ 1 \end{pmatrix}$ the operator $b_1^\dagger$ reads as a 2×2 matrix in spin space with bosonic operator valued entries

$$b_1^\dagger = (ic)^2 \begin{pmatrix} -a^2 \frac{\beta^2}{(ic)^2} e^{-2\Theta^-} + a \frac{e^{-\Theta^-}\beta}{\kappa ic}\left(-\frac{1}{2} - \delta_b - \tilde{\xi} + n\right) + s^2 + (\delta_b - n)^2 & -2a \frac{\beta}{ic} \\ -2a e^{-2\Theta^-} \frac{\beta}{ic} + \frac{e^{-\Theta^-}}{\kappa}\left(\frac{1}{2} - s - \tilde{\xi}\right) \\ -a^2 e^{-2\Theta^-} \frac{\beta^2}{(ic)^2} + \frac{e^{-\Theta^-}\beta}{\kappa ic}\left(\frac{3}{2} - \delta_b - \tilde{\xi} + n\right) a + (s-1)^2 + (\delta_b - n)^2 \end{pmatrix} \tag{4.58}$$

where $n = a^\dagger a$ is the bosonic number operator, $\tilde{\xi} \equiv \xi/(ic)$ and the lines are for guidance to the eye. This operator will act on some product state of bosons and spin

$$|\psi\rangle = \sum_{n\sigma} \psi_{n\sigma} |n\sigma\rangle = \sum_{n=0}^{\infty} \left(\psi_{n\uparrow}|n\rangle \otimes |\uparrow\rangle + \psi_{n\downarrow}|n\rangle \otimes |\downarrow\rangle\right)$$
$$= \sum_{n=0}^{\infty} \begin{pmatrix} \psi_{n\uparrow}|n\rangle \\ \psi_{n\downarrow}|n\rangle \end{pmatrix} \tag{4.59}$$

leading to the eigenvalue problem

$$\left(b_1^\dagger - E \mathbb{1}\right)|\psi\rangle = 0 \tag{4.60}$$

with eigenvalue E and the identity operator $\mathbb{1}$.

As the set of bosonic states is orthogonal the coefficient for each $|n\rangle$ has to vanish separately leading to two intertwined recursion relations for the coefficients $\psi_{n\sigma}$ of the eigenstate

$$0 = -\sqrt{n+2}\sqrt{n+1} e^{-2\Theta^-} \psi_{n+2\downarrow} - \sqrt{n+1} \frac{e^{-\Theta^-}}{\kappa} \left((\delta_b - n) + (\tilde{\xi} + \tfrac{1}{2})\right)\psi_{n+1\uparrow}$$
$$+ \left(s^2 + (\delta_b - n)^2 - \tilde{E}\right)\psi_{n\uparrow} - 2\sqrt{n+1} e^{-2\Theta^-} \psi_{n+1\downarrow} + \frac{e^{-\Theta^-}}{\kappa}\left(\tfrac{1}{2} - s - \tilde{\xi}\right)\psi_{n\downarrow} \tag{4.61}$$

$$0 = -2\sqrt{n+1}\psi_{n+1\uparrow} - \sqrt{n+2n+1} e^{-2\Theta^-}\psi_{n+2\downarrow} + \sqrt{n+1}\frac{e^{-\Theta^-}}{\kappa}$$
$$\times \left(-(\delta_b - (n+1)) + (\tilde{\xi} - \tfrac{1}{2})\right)\psi_{n+1\downarrow} + \left((s-1)^2 + (\delta_b - n)^2 - \tilde{E}\right)\psi_{n\downarrow} \,. \tag{4.62}$$

51

4. Separation of Variables

Here we have scaled the eigenvalue using $\widetilde{E} \equiv E/(ic)^2$ and again $\tilde{\xi} = \xi/(ic)$. Considering the large n regime (in comparison to the corresponding eigenvalues) each coefficient satisfies functional relation similar to Γ-function relations. Thus to obtain normalizable states the recursion relation needs to terminate at some finite value m of bosons.

For this case the eigenvalue can be read off directly from the coefficient of this state $|m\rangle$ with the highest number of bosons in (4.60) leading to

$$\begin{pmatrix} \left((\delta_b - m)^2 + s^2 - \widetilde{E}\right)\psi_{m\uparrow} + \left(\frac{e^{-\Theta^-}}{\kappa}(\frac{1}{2} - s - \tilde{\xi})\right)\psi_{m\downarrow} \\ \left((\delta_b - m)^2 + (s-1)^2 - \widetilde{E}\right)\psi_{m\downarrow} \end{pmatrix} |m\rangle = 0. \quad (4.63)$$

There are two possibilities to satisfy the set of equations (4.63)

$$\widetilde{E}_1^{b_1} = (\delta_b - m)^2 + s^2 \quad , \qquad \psi_{m\downarrow} = 0 \quad , \quad \psi_{m\uparrow} \neq 0 \text{ arbitrary} \quad (4.64)$$

$$\widetilde{E}_2^{b_1} = (\delta_b - m)^2 + (s-1)^2 \quad , \qquad \frac{\psi_{m\uparrow}}{\psi_{m\downarrow}} = \frac{e^{-\Theta}}{\kappa} \frac{\frac{1}{2} - s - \tilde{\xi}}{1 - 2s}. \quad (4.65)$$

The corresponding eigenstates are then calculated by explicitly carrying out the recursions (4.61) and (4.62) with $\psi_{m+1,0} = 0$ and the values of $\psi_{m,0}$ stated above.

Turning to the operator b_2 an analogous calculation results in

$$\bar{E}_1^{b_2} = (\delta_b - m)^2 s^2 \quad , \qquad \psi_{m\downarrow} = 0 \quad , \quad \psi_{m\uparrow} \neq 0 \text{ arbitrary} \quad (4.66)$$

$$\bar{E}_2^{b_2} = (\delta_b - m)^2 (s-1)^2 \quad , \qquad \frac{\psi_{m\uparrow}}{\psi_{m\downarrow}} = \frac{e^{-\Theta}}{\kappa} \frac{\frac{1}{2} - s - \tilde{\xi}}{1 - 2s} \quad (4.67)$$

where we introduced $\bar{E} \equiv E/(ic)^4$. The eigenstates again result from analogous recursion relations.

As b_1 and b_2 commute due to $[\mathscr{B}(\lambda), \mathscr{B}(\mu)] = 0$ they share a common system of eigenvectors. The respective eigenvalues of $b_1^\dagger$ or $b_2^\dagger$ are only degenerate for a finite number of states if the inhomogeneities δ_s and δ_b are chosen carefully. In such cases this degeneracy is lifted by $b_2^\dagger$ or $b_1^\dagger$ respectively. The matter of degeneracy is delicate as there exist some pathologic scenarios where the method of separation of variables fails. E.g. $s = \frac{1}{2}$ results in massive degeneracy of b_1 and b_2 or some integer values of δ_b lead to degeneracy of a finite number of states which then have to be considered particularly.

Example. For $s = 0$ and $\delta_b = 0$ only the states for $m = 0$, $\psi_{0\downarrow} \neq 0$, $\psi_{0\uparrow} \neq 0$ and $m = 1$, $\psi_{1\uparrow} \neq 0$, $\psi_{1\downarrow} = 0$ are degenerate regarding the operator $b_1^\dagger$. Hence to be sure of the right-eigenbasis it only remains to check if the right-eigenstates of $b_1^\dagger$ for these cases are eigenstates of $b_2^\dagger$ which is trivially fulfilled.

With the eigenvalues at hand it is possible to write the operator valued zeros of $\mathscr{B}$ as a matrix acting as multiplication operators on the common eigenbasis of $b_1^\dagger$ and $b_2^\dagger$ from (4.57). In each sector of bosonic numbers m we find

$$\hat{x}_b^2 = (ic)^2 \begin{pmatrix} (\delta_b - m)^2 & 0 \\ 0 & (\delta_b - m)^2 \end{pmatrix} \quad , \quad \hat{x}_s^2 = (ic)^2 \begin{pmatrix} s^2 & 0 \\ 0 & (s-1)^2 \end{pmatrix} \quad (4.68)$$

4.3. Spin Boson Model

As the $\hat{x}^2$ operators can be simultaneously diagonalized (c.f. (4.8)) the 'coordinates' are

$$\hat{x}_b = \mathrm{ic}\begin{pmatrix} \delta_b - m & 0 \\ 0 & \delta_b - m \end{pmatrix} \quad , \quad \hat{x}_s = \mathrm{ic}\begin{pmatrix} \delta_s + \frac{1}{2} & 0 \\ 0 & \delta_s - \frac{1}{2} \end{pmatrix} \tag{4.69}$$

This also fixes the sets Λ for the lattice of the TQ-equation

$$\Lambda_b = \{\ldots, \mathrm{ic}(\delta_b - 2), \mathrm{ic}(\delta_b - 1), \mathrm{ic}\delta_b\} \quad , \quad \Lambda_s = \{\mathrm{ic}(\delta_s - \tfrac{1}{2}), \mathrm{ic}(\delta_s + \tfrac{1}{2})\} \tag{4.70}$$

Using the operator-valued zeros we define the 'conjugated momenta' to the 'coordinates' $\hat{x}_j$ with $j \in \{s, b\}$ analogously to the spin chain scenario as

$$\left._{\lambda = \hat{x}_j}\middle|\mathscr{A}(\lambda) = \sum_p \hat{x}_j^p \mathscr{A}_p \equiv X_j^-\right. \\ \left._{\lambda = \hat{x}_j}\middle|\widetilde{\mathscr{D}}(\lambda) = \sum_p \hat{x}_j^p \widetilde{\mathscr{D}}_p \equiv X_j^+\right. \tag{4.71}$$

These operators were found to act as $X_j^\pm = \Delta_j^\pm E_j^\pm$ involving the factorization of the quantum determinant $\Delta_q(\lambda) = \Delta^-(\lambda + \frac{\mathrm{ic}}{2})\Delta^+(\lambda - \frac{\mathrm{ic}}{2})$ and the shift operators on the grid $E_j^\pm$

$$E_s^\pm : \mathbb{C}^2 \to \mathbb{C}^2 : (x_s, x_b) \mapsto (x_s \pm \mathrm{ic}, x_b) \\ E_b^\pm : \mathbb{C}^2 \to \mathbb{C}^2 : (x_s, x_b) \mapsto (x_s, x_b \pm \mathrm{ic}) . \tag{4.72}$$

Due to the algebra generated by $\{\hat{x}_s, \hat{x}_b, X_s^\pm X_b^\pm\}$ the factorization of the quantum determinant needs to obey (4.31) where now $n, m \in \{s, b\}$ only. The following factorization meets these demands

$$\Delta^-(\lambda) = \frac{\lambda - \frac{\mathrm{ic}}{2} + \alpha^-}{\alpha^-}(\lambda + s\mathrm{ic})(\lambda - (s-1)\mathrm{ic})(\frac{\beta\gamma}{\mathrm{ic}}(\lambda + \delta_b\mathrm{ic})) \\ \Delta^+(\lambda) = (2\lambda - \mathrm{ic})\frac{\lambda + \frac{\mathrm{ic}}{2} - \alpha^-}{\alpha^-}(\lambda + (s-1)\mathrm{ic})(\lambda - s\mathrm{ic})(\frac{\beta\gamma}{\mathrm{ic}}(\lambda - \delta_b\mathrm{ic})) \tag{4.73}$$

and $\Delta^\pm$ vanish on the appropriate boundaries of the sets $\Lambda_{s,b}$. Notice that the set Λ_b is 'half' infinite and hence only $\Delta^+(\delta_b\mathrm{ic})$ is demanded to vanish. Further note that we used the parametrization (2.30) of the boundary matrix again to obtain suitable equations. The lattice and the factorization at hand, we arrive at the TQ-equation by considering the spectral problem as in section 4.1.5

$$\Lambda(x_j)Q(x_j) = \frac{(x_j + \frac{\mathrm{ic}}{2})(x_j + \alpha^+ - \frac{\mathrm{ic}}{2})}{2x_j\alpha^+}\Delta^-(x_j)Q(x_j - \mathrm{ic}) \\ - \frac{x_j - \alpha^+ + \frac{\mathrm{ic}}{2}}{4x_j\alpha^+}\Delta^+(x_j)Q(x_j + \mathrm{ic}) \tag{4.74}$$

with an unknown function $Q(x)$. The allowed arguments are $x_j \in \Lambda_j$ on the grid entering $\Delta_j^\pm(\mathbf{x}) = \Delta^\pm(x_j)$ for $j \in \{s, b\}$.

53

4. Separation of Variables

The vanishing of $\Delta^{\pm}$ on the boundaries of $\Lambda_s = \{x_s^- \equiv ic(\delta_s - \frac{1}{2}) = (s-1)ic, x_s^+ \equiv ic(\delta_s + \frac{1}{2}) = sic\}$ yields

$$\Lambda(x_s^+)Q(x_s^+) = \frac{(x_s^+ + \frac{ic}{2})(x_s^+ + \alpha^+ - \frac{ic}{2})}{2x_s^+\alpha^+}\Delta^-(x_s^+)Q(x_s^-)$$
$$\Lambda(x_s^-)Q(x_s^-) = -\frac{x_s^- - \alpha^+ + \frac{ic}{2}}{4x_s^-\alpha^+}\Delta^+(x_s^-)Q(x_s^+)$$
(4.75)

leading to a determinant of a 2×2 matrix being zero for a non-trivial solution. And the vanishing of Δ^+ at the 'upper' boundary $\delta_b ic \equiv x_b^0$ of $\Lambda_b = \{\ldots, \delta_b ic - n\,ic, \ldots, \delta_b ic - ic, \delta_b ic\} \equiv \{\ldots, x_b^{-n}, \ldots, x_b^{-1}, x_b^0\}$ results for the boundary in

$$\Lambda(x_b^0)Q(x_b^0) = \frac{(x_b^0 + \frac{ic}{2})(x_b^0 + \alpha^+ - \frac{ic}{2})}{2x_b^0\alpha^+}\Delta^-(x_b^0)Q(x_b^{-1})$$
(4.76)

and for all other lattice points of Λ_b we have with $n > 0$

$$\Lambda(x_b^{-n})Q(x_b^{-n}) = \frac{(x_b^{-n} + \frac{ic}{2})(x_b^{-n} + \alpha^+ - \frac{ic}{2})}{2x_b^{-n}\alpha^+}\Delta^-(x_b^{-n})Q(x_b^{-(n+1)})$$
$$- \frac{x_b^{-n} - \alpha^+ + \frac{ic}{2}}{4x_b^{-n}\alpha^+}\Delta^+(x_b^{-n})Q(x_b^{-(n-1)}).$$
(4.77)

The relations (4.76) and (4.77) lead to a continuant[4] of a half-infinite matrix being zero.

Besides the easily obtainable asymptotic the expansion of eigenvalue Λ in powers of the spectral parameter λ involves three other coefficients a_j

$$\Lambda(\lambda) = \frac{2e^{\theta^- - \theta^+}}{\alpha^+\alpha^- \operatorname{ch}\beta^+ \operatorname{ch}\beta^-}\lambda^6 + a_4\lambda^4 + a_2\lambda^2 + a_0.$$
(4.78)

In addition to the vanishing conditions of the 2×2 determinant and of the half-infinite continuant we have a third equation from the c-number valued zero $ic/2$ of the $\mathscr{B}$ operator resulting again in $\Lambda(0) = (d_q T)(-ic/2)$. Hence the eigenvalue is in principle fully obtainable but unfortunately the condition of the vanishing half-infinite continuant has not been found to be evaluatable and states an open problem of its own.

The TQ-equation for the known cases of constricted boundary parameters in [5] holds for general arguments. In this spirit we are now leaving the grid behind and consider (4.74) for a continuous argument, i.e.

$$\Lambda(\lambda)Q(\lambda) = \tilde{\Delta}^-(\lambda)Q(\lambda - ic) + \tilde{\Delta}^+(\lambda)Q(\lambda + ic).$$
(4.79)

---
[4] A *continuant* is a determinant of a matrix all of whose elements are zero except those in the main diagonal and in the two adjacent diagonal lines parallel to and on either side of the main diagonal [70].

4.3. Spin Boson Model

One finds that the coefficients

$$\begin{aligned}
\tilde{\Delta}^-(\lambda) &= \frac{(\lambda + ic/2)(\lambda + \alpha^+ - ic/2)}{2\lambda \alpha^+} \Delta^-(x_j) \\
&= \frac{\beta \gamma}{2ic\alpha^+ \alpha^-} \frac{1}{\lambda} \left(\lambda + \frac{ic}{2}\right) \left(\lambda + \alpha^+ - \frac{ic}{2}\right) \left(\lambda + \alpha^- - \frac{ic}{2}\right) \times \\
&\quad \times (\lambda + s\,ic)(\lambda - (s-1)ic)(\lambda - \delta_b ic) \sim \lambda^5
\end{aligned} \qquad (4.80)$$

$$\tilde{\Delta}^+(\lambda) = \tilde{\Delta}^+(-\lambda) \sim -\lambda^5$$

behave asymptotically as λ^5 for large values of the spectral parameter. Hence the mismatch in the asymptotic behaviour of the left and right hand side in (4.79) must be compensated by the Q-function. This is possible by some Γ-function dependence of the Q-function but the explicit form of the function remains an unresolved task.

Note that for Bethe ansatz solvable cases, i.e. diagonal or triangular boundary matrices, the coefficient of λ^6 in (4.78) vanishes and the asymptotical behaviour is $\sim \lambda^4$. Furthermore the λ^5 coefficient on the right hand side of (4.79) cancels and the TQ-equation can by solved using an even polynomial as an ansatz for $Q(\lambda)$ in agreement with the Bethe ansatz analysis carried out in [5].

Chapter 5

XXZ Model with Twisted Boundary Conditions

In this chapter we want to pick up the question of relations and differences of the method of separation of variables to the approach using the fusion hierarchy applied to the open boundary XXZ chain by Murgan, Nepomechie *et al.* which arose in section 4.2. To this end we need to consider a slightly simpler setting where both methods are applicable for the solution of the spectral problem: The XXZ model with general toroidal boundary conditions is defined by the hamiltonian

$$\mathcal{H} = \sum_{j=1}^{L} \left[\sigma_j^x \sigma_{j+1}^x + \sigma_j^y \sigma_{j+1}^y + \operatorname{ch} \eta\, \sigma_j^z \sigma_{j+1}^z \right], \qquad \sigma_{L+1}^\alpha = K^{-1} \sigma_1^\alpha K \tag{5.1}$$

where σ_j^α, $\alpha = x, y, z$ denote the Pauli matrices for spins-$\frac{1}{2}$ at site j. The unitary matrix $K \in \operatorname{End}(\mathbb{C}^2)$ determines the boundary conditions. For anti-diagonal K the model is integrable but has no pseudo vacuum state. The spectral problem has first been solved by means of Baxter's method [8] and solutions to the resulting functional equations can be given in terms of the roots of Bethe ansatz equations. In the following we will first review this solution and the method of separation of variables very briefly for this model [77]. Then we present the solution using the approach to the spectral problem based on the fusion hierarchy of transfer matrices at anisotropies $\eta = \mathrm{i}\pi/(p+1)$ with integer $p > 1$.

The transfer matrix $\hat{t}$ generating the hamiltonian (5.1) differs from the transfer matrix for the periodic model (2.17) only by the boundary matrix K which needs to a be a representation of the Yang-Baxter algebra (2.1) itself

$$\hat{t}(\lambda) = \operatorname{tr}(K T(\lambda)). \tag{5.2}$$

5. XXZ Model with Twisted Boundary Conditions

For the R-matrix with entries (2.4) the Yang-Baxter algebra has two classes of constant representations namely diagonal or anti-diagonal twist matrices K. Without loss of generality [2, 100] the restriction to

$$K = \begin{pmatrix} e^{-i\Phi} & 0 \\ 0 & e^{i\Phi} \end{pmatrix} \quad , \quad K = \begin{pmatrix} 0 & 1 \\ 1 & 0 \end{pmatrix} \tag{5.3}$$

with a twist angle Φ in the diagonal case is possible. The periodic model is recovered for $\Phi = 0$. For this model we want to refer to $KT(\lambda)$ as the monodromy matrix for convenience. Again it has quantum space operators as entries $A(\lambda), B(\lambda), C(\lambda)$ and $D(\lambda)$. Applying a logarithmic derivative to the transfer matrix (5.2) reveals the spin chain hamiltonian (5.1) to be a member of the generated family of commuting operators

$$H = 2\sh\eta \left.\frac{\partial \ln \hat{t}(\lambda)}{\partial \lambda}\right|_{\lambda=\frac{\eta}{2}} - L\ch\eta . \tag{5.4}$$

5.1 Baxter's Method and Separation of Variables

For a diagonal twist matrix the eigenvalues and eigenstates of the transfer matrix $t(\lambda)$ can be obtained by means of the algebraic Bethe ansatz starting from the ferromagnetic so-called pseudo vacuum with polarized spins (see e.g. [62]) and will not be discussed any further. For the anti-diagonal twist matrix

$$K = \begin{pmatrix} 0 & 1 \\ 1 & 0 \end{pmatrix}, \tag{5.5}$$

however, the total magnetization is not a good quantum number. As a consequence there is no simple reference state such as the ferromagnetic one and the algebraic Bethe ansatz cannot be applied. Instead, a TQ-equation for the eigenvalues $\hat{\Lambda}(\lambda)$ of the transfer matrix $\hat{t}(\lambda)$ has been obtained using Baxter's method of commuting transfer matrices [8, 100]

$$\hat{\Lambda}(\lambda)q(\lambda) = \sh^L\left(\lambda + \tfrac{\eta}{2}\right)q(\lambda - \eta) - \sh^L\left(\lambda - \tfrac{\eta}{2}\right)q(\lambda + \eta) . \tag{5.6}$$

This difference equation is solved by

$$q(\lambda) = \prod_{j=1}^{L} \sh\tfrac{1}{2}(\lambda - \lambda_j) . \tag{5.7}$$

As a consequence of the analyticity of the transfer matrix eigenvalues it follows that the rapidities λ_j are pairwise different solutions to the Bethe equations [1]

$$\frac{\sh^L(\lambda + \tfrac{\eta}{2})}{\sh^L(\lambda - \tfrac{\eta}{2})} = -\prod_{k\neq j}\frac{\sh\tfrac{1}{2}(\lambda - \lambda_k + \eta)}{\sh\tfrac{1}{2}(\lambda - \lambda_k - \eta)} \quad , \quad \text{for } \lambda \in \{\lambda_j\}_{j=1}^{L} . \tag{5.8}$$

---
[1] Note that these equations with an extra phase (-1) and any number $M \leq L$ of rapidities λ_j determine the spectrum of a staggered six-vertex model [43]. This case, however, cannot be obtained from (5.1) with the twist matrices (5.3).

5.1. Baxter's Method and Separation of Variables

From (5.4) we obtain the corresponding eigenvalue of the spin chain hamiltonian (5.1)

$$E(\{\lambda_j\}) = L\,\text{ch}\,\eta + 2\sum_{j=1}^{L} \frac{\text{sh}\,\eta\,\text{sh}\,\frac{\eta}{2}}{\text{ch}\,\lambda_j - \text{ch}\,\frac{\eta}{2}}. \tag{5.9}$$

The method of separation of variables for this model, carried out in [77, 78], is only applicable to the anti-diagonal twist matrix as in the diagonal case the B operator in the monodromy matrix is not a polynomial of maximal degree. Inserting inhomogeneities s_j at each lattice site the factorization with operator valued zeros $\hat{x}_j$ is

$$B(\lambda) = \text{sh}(\lambda - \hat{x}_1)\cdots\text{sh}(\lambda - \hat{x}_L), \tag{5.10}$$

where $\hat{x}_j = \text{diag}(x_j^1, \ldots, x_j^{2^L})$ in the eigenbasis of $B(\lambda)$ and $x_j^\ell = s_j \pm \frac{\eta}{2}$ forming the sets Λ_j. Introducing the 'conjugated' momenta the needed commutation relations hold and we finally arrive at (4.31) again with

$$\Delta_j^\pm(\mathbf{x}) = \Delta_\pm(x_j), \tag{5.11}$$

where

$$\Delta_\pm(\lambda) = \xi_\pm \text{sh}(\lambda - s_1 \mp \tfrac{\eta}{2})\cdots\text{sh}(\lambda - s_L \mp \tfrac{\eta}{2}). \tag{5.12}$$

In this definition the constants $\xi_\pm$ are an arbitrary factorization of the determinant of the twist matrix

$$\xi_+\xi_- = \det(K) = -1. \tag{5.13}$$

The functions $\Delta_\pm$ factorize the quantum determinant of the monodromy matrix $(d_qT)(\lambda - \frac{\eta}{2}) = -\left[\prod_{j=1}^{L} \text{sh}(\lambda - s_j + \eta)\text{sh}(\lambda - s_j - \eta)\right]$ in the following sense

$$\Delta_+(s_j - \tfrac{\eta}{2})\Delta_-(s_j + \tfrac{\eta}{2}) = (d_qT)(s_j - \tfrac{\eta}{2}). \tag{5.14}$$

Turning to the spectral problem as in section 4.1.5 and 4.2 we find

$$\hat{\Lambda}(x_j)q_j(x_j) = \Delta_+(x_j)q_j(x_j + \eta) + \Delta_-(x_j)q_j(x_j - \eta), \qquad j = 1, \ldots, L, \tag{5.15}$$

which we recognize as the TQ-equation (5.6) evaluated on a discrete lattice $x_j \in \Lambda_j$.

Again the solution for small system sizes can be obtained by eliminating q from the equations (5.15) analogous to section 4.1.5

$$\hat{\Lambda}(s_j + \tfrac{\eta}{2})\hat{\Lambda}(s_j - \tfrac{\eta}{2}) = \Delta(s_j), \qquad j = 1, \ldots, L \tag{5.16}$$

and starting with a polynomial ansatz for the eigenvalue

$$\hat{\Lambda}(\lambda) = a_{-L+1}e^{(-L+1)\lambda} + a_{-L+3}e^{(-L+3)\lambda} + \cdots + a_{L-1}e^{(L-1)\lambda}. \tag{5.17}$$

The L coefficients a_j have to be determined from the L equations (5.16).

5.2 Fusion Hierarchy and Truncation Identity at Roots of Unity

This method was first developed for the restricted solid-on-solid model (RSOS) by Bazhanov et al. [10] and was adapted to spin chains by Nepomechie e.g. [74, 75]. Unfortunately this method only works if the values of the crossing parameter are chosen to be roots of unity $\eta = i\pi/(p+1)$. Nevertheless like in the periodic case [75] the solution obtained is valid for arbitrary η as it coincides with (5.6).

We have demonstrated in section 2.3 how the fusion procedure in the auxiliary space allows to construct a fusion hierarchy for higher spin transfer matrices (2.73) associated to the periodic XXZ model. As the twist matrices K (5.3) are a representation of the Yang-Baxter algebra as well their fusion procedure is described analogously to (2.63) resulting in the spin-$\frac{k}{2}$ boundary matrix

$$K_{\langle 1 \cdots k\rangle} = P^+_{1 \cdots k} K_1 K_2 \cdots K_k P^+_{1 \cdots k}. \tag{5.18}$$

As a consequence the same fusion hierarchy (2.73) holds for the transfer matrix (5.2) as for the transfer matrix (2.17).

On the other hand fused transfer matrices can be constructed using quantum-group theory [63, 74]. The R-matrices of higher auxiliary spins from quantum-group constructions have a simple direct relation to an R-matrix with lower auxiliary spin at roots of unity $\eta = i\pi/(p+1)$ with p being an integer number. It is also possible to relate the quantum-group R-matrices to those constructed via fusion. Resulting in the identity at roots of unity

$$B_{1 \cdots p+1} A_{1 \cdots p+1} R_{\langle 1 \cdots p+1\rangle, p+2}(\lambda) A^{-1}_{1 \cdots p+1} B^{-1}_{1 \cdots p+1} =$$

$$\mu(\lambda) \begin{pmatrix} \nu(\lambda)\sigma^z & & \\ & B_{1 \cdots p-1} A_{1 \cdots p-1} R_{\langle 1 \cdots p-1\rangle, p}(\lambda+\eta) A^{-1}_{1 \cdots p-1} B^{-1}_{1 \cdots p-1} & \\ & & -\nu(\lambda)\sigma^z \end{pmatrix} \tag{5.19}$$

where the entries of matrix A are unnormalized Clebsch-Gordon coefficients in the decomposition of the tensor product of k spin-1/2 representations into a direct sum of SU(2) irreducible representations and the matrix B is a diagonal matrix needed for symmetrizing (see appendix C or [74]).

The function $\mu(\lambda)$ is related to the quantum determinant

$$(d_q T)(\lambda - \eta) = -\left(-\mu(\lambda)\right)^L \tag{5.20}$$

$$\mu(\lambda) \equiv -\operatorname{sh}(\lambda + \tfrac{\eta}{2}) \operatorname{sh}(\lambda - \tfrac{3\eta}{2}) \tag{5.21}$$

and $\nu(\lambda) \equiv -\mu(\lambda)^{-1} \left(\tfrac{i}{2}\right)^p \operatorname{sh}\left((p+1)(\lambda - \tfrac{\eta}{2})\right)$ is related to the crossing parameter via p.

The fused twist matrices themselves obey a truncation identity similar to (5.19). Under the fusion procedure an anti-diagonal matrix with only 1's as entries remains anti-diagonal

5.2. Fusion Hierarchy and Truncation Identity at Roots of Unity

with 1's as entries after applying the transformation of the appropriate Clebsch-Gordon matrix and omitting null rows and columns, hence

$$A_{1\cdots k}K_{(1\cdots k)}A^{-1}_{1\cdots k} = \begin{pmatrix} & & 1 \\ & A_{1\cdots k-1}K_{(1\cdots k-1)}A^{-1}_{1\cdots k-1} & \\ 1 & & \end{pmatrix}. \tag{5.22}$$

Identities (5.19) and (5.22) together give a truncation identity for the product of twist and monodromy matrix

$$B_{1\cdots p+1}A_{1\cdots p+1}KT_{(1\cdots p+1),p+2}(\lambda)A^{-1}_{1\cdots p+1}B^{-1}_{1\cdots p+1} = \mu(\lambda)^L \times$$
$$\begin{pmatrix} & & (-\nu(\lambda))^L F \\ & B_{1\cdots p-1}A_{1\cdots p-1}KT_{(1\cdots p-1),p}(\lambda+\eta)A^{-1}_{1\cdots p-1}B^{-1}_{1\cdots p-1} & \\ \nu(\lambda)^L F & & \end{pmatrix} \tag{5.23}$$

with $F \equiv \prod_{j=1}^{L} \sigma_j^z$ accordingly for the transfer matrix we find by taking the trace of (5.23)

$$\hat{t}_{(p+1)}(\lambda) = -(d_q T)(\lambda - \eta)(-1)^L \hat{t}_{(p-1)}(\lambda + \eta). \tag{5.24}$$

The fusion hierarchy (2.73) together with the truncation identity (5.24) leads to a functional relation for the transfer matrix at roots of unity for a given p, e.g. for $p = 2$ or $k = 2$ respectively this relation is

$$\hat{t}(\lambda)\hat{t}(\lambda+\eta)\hat{t}(\lambda+2\eta) - (d_q T)(\lambda)\hat{t}(\lambda+2\eta) + (d_q T)(\lambda+\eta)\hat{t}(\lambda) + $$
$$+ (-1)^L (d_q T)(\lambda-\eta)\hat{t}(\lambda+\eta) = 0. \tag{5.25}$$

Like in the RSOS model [10] or the periodic XXZ chain [75] the goal is to recast the general form of the functional relation (5.25) as a determinant of a certain matrix. This determinant being zero ensures the existence of a null eigenvector which leads to equations similar to a TQ-equation.

In the case of an anti-diagonal K-matrix the functional relation found above cannot be recast directly, though multiplying it with itself shifted by $i\pi = (p+1)\eta$ results in a recastable expression. For general p this is a determinant of a $(2p+2) \times (2p+2)$ matrix reading

$$\det \begin{pmatrix} \hat{\Lambda}_0 & h_0 & 0 & \cdots & 0 & -(-1)^N h_1 \\ -h_2 & \hat{\Lambda}_1 & h_1 & & & 0 \\ 0 & -h_3 & \hat{\Lambda}_2 & \ddots & & \vdots \\ \vdots & & \ddots & \ddots & & 0 \\ 0 & & & -h_{2p+1} & \hat{\Lambda}_{2p} & h_{2p} \\ (-1)^N h_{2p+1} & 0 & \cdots & 0 & -h_0 & \hat{\Lambda}_{2p+1} \end{pmatrix} = 0 \tag{5.26}$$

5. XXZ Model with Twisted Boundary Conditions

with the eigenvalue $\hat{\Lambda}$ of the transfer matrix $\hat{t}$. In the above expression we used the shorthands

$$\hat{\Lambda}_k \equiv \hat{\Lambda}(\lambda + k\eta) \qquad (5.27)$$

$$h_k \equiv \text{sh}^L(\lambda + k\eta - \tfrac{\eta}{2}). \qquad (5.28)$$

The definition of h_k directly reveals $h_k = (-1)^L h_{p+1+k}$. This and the periodicity of the eigenvalue $\hat{\Lambda}_k = -(-1)^L \hat{\Lambda}_{p+1+k}$, following from

$$R_j(\lambda + i\pi) = -\sigma_0^z R_j(\lambda)\sigma_0^z = -\sigma_j^z R_j(\lambda)\sigma_j^z \qquad (5.29)$$

where the subscript 0 denotes the auxiliary space, are needed to verify the equivalence of the determinant and the product of functional relations.

Let $(q_0, q_1, \ldots, q_{2p+1})$ be the null eigenvector of the matrix, this yields the equations

$$\begin{aligned}
&\hat{\Lambda}_0 q_0 + h_0 q_1 - (-1)^L h_1 q_{2p+1} = 0 \\
&-h_{k+1} q_{k-1} + \hat{\Lambda}_k q_k + h_k q_{k+1} = 0 \quad \text{for } k = 1, \ldots, 2p \\
&(-1)^L h_{2p+1} q_0 - h_0 q_{2p} + \hat{\Lambda}_{2p+1} q_{2p+1} = 0.
\end{aligned} \qquad (5.30)$$

Using the ansatz $q_k = q(\lambda + k\eta)$ with

$$q(\lambda) = \prod_{j=1}^{L} \text{sh}\tfrac{1}{2}(\lambda - \lambda_j) \qquad (5.31)$$

the equations (5.30) imply only a single TQ-equation

$$\hat{\Lambda}(\lambda)q(\lambda) = \text{sh}^L(\lambda + \tfrac{\eta}{2}) q(\lambda - \eta) - \text{sh}^L(\lambda - \tfrac{\eta}{2}) q(\lambda + \eta) \qquad (5.32)$$

agreeing with (5.6) and leading to the same Bethe ansatz equations (5.8). Notice the $2\pi i$ periodicity of the q-function arising from the $2(p+1)$ rows of the matrix in (5.26) and the product in (5.31) running up to L due to the structure of the upper right and lower left entries.

Both methods, i.e. the method of separation of variables and the fusion approach, presented here arrive at same TQ-equation (5.6). With the first method it has to be solved on a lattice of singular points of this functional equation only, while the latter approach only allows for anisotropies being roots of unity. This is unlike the situation for the spin chain with open boundaries and non-diagonal boundary fields where different functional equations have been found within different approaches. Hence we conclude that it is not a general feature of the methods but rather a peculiarity of the respective model. In the situation of the open XXX chain (chapter 4) compared to the open XXZ chain [71, 73] the difference of the number and form of the TQ-equations might be a consequence of the rational limit which is not included in the approach presuming anisotropies of roots of unity.

Chapter 6

Non-linear Integral Equations for the XXX Spin Chain

In this chapter we want to continue the pursuit of the correct eigenvalues of the XXX spin chain hamiltonian with non-diagonal boundary fields (1.2). With the functional Bethe ansatz applied to the model in chapter 4 we were just able to obtain correct finite size corrections for small chain lengths as the determining equations were only valid on a lattice and a laborious multidimensional root finding needed to be carried out. The missing off-lattice corrections to functional equations (4.40) are included in the fusion hierarchy (2.85) hence it is promising to try to determine the finite size corrections from it.

In the following sections we derive two different kinds of non-linear integral equations (NLIE) from the fusion hierarchy (2.85). In section 6.1 we will follow [89] to derive a single integral equation valid for all states of the model while in section 6.2 we derive an infinite set of equations with driving terms governed by the particularly chosen state. The latter approach is numerically treatable and provides the eigenvalue for certain parameter regimes.

6.1 Single Non-linear Integral Equation

The construction of the single NLIE requires only the first level of the fusion hierarchy (2.85). In this section we take without loss of generality $c = i$ and introduce inhomogeneities s_j for each lattice site j like in example containing equation (2.24). Hence the first equation reads (2.84) reads in the rational limit

$$t_2(\lambda) = t_1(\lambda + 1)t_1(\lambda) - \Delta(\lambda + 1) \qquad (6.1)$$

6. Non-linear Integral Equations for the XXX Spin Chain

with $\Delta(u)$ being the rational limit of (2.83)

$$\Delta(\lambda) = \frac{\lambda^2 - 1}{4\lambda^2 - 1}\left[\prod_{\ell=1}^{L}(\lambda - s_\ell - 1)(\lambda - s_\ell + 1)(\lambda + s_\ell - 1)(\lambda + s_\ell + 1)\right]B(\lambda) \qquad (6.2)$$

mainly resembling the quantum determinant of the model. The explicit dependence of the diagonal boundary parameters is collected in the function

$$B(\lambda) = \frac{(\lambda^2 - \alpha_-^2)(\lambda^2 - \alpha_+^2)}{(\alpha_- \alpha_+)^2}. \qquad (6.3)$$

As the transfer matrix generating the hamiltonian is a polynomial in the spectral parameter its eigenvalue t_1 is a polynomial as well. Since $\Delta(\lambda)$ is a non-polynomial rational function all eigenvalues of the fused transfer matrices t_j for $j \neq 1$ are not polynomial. Nevertheless certain zeros and poles of the fused transfer matrices can easily be identified numerically and removed by introducing

$$\tilde{t}_k(\lambda) = \frac{\prod_{m=0}^{k-1}(\lambda + m + \frac{1}{2})}{\prod_{j=1}^{k-1}(\lambda + j)\left(\prod_{\ell=1}^{L}(\lambda - s_\ell + j)(\lambda + s_\ell + j)\right)} t_k(\lambda). \qquad (6.4)$$

Especially after this transformation we find $\tilde{t}_2$ to be polynomial

$$\tilde{t}_2(\lambda) = \frac{(\lambda + \frac{3}{2})(\lambda + \frac{1}{2})}{(\lambda + 1)\left(\prod_{\ell=1}^{L}(\lambda - s_\ell + 1)(\lambda + s_\ell + 1)\right)} t_2(\lambda) \qquad (6.5)$$

because of (2.85) and $\tilde{t}_1$ remaining polynomial. As we want to use Cauchy's integral formula to derive the integral equation we insert singularities at controlled positions via

$$f_1(\lambda) = \frac{\tilde{t}_1(\lambda)}{\psi(\lambda)} \qquad (6.6)$$

$$\psi(\lambda) = \left[\prod_{\ell=1}^{L}(\lambda - s_\ell)(\lambda + s_\ell + 1)\right]\lambda(\lambda + \tfrac{1}{2})(\lambda + 1) \qquad (6.7)$$

leaving us with the key relation

$$f_1(\lambda + \tfrac{1}{2})f_1(\lambda - \tfrac{1}{2}) = f_2(\lambda) + b(\lambda). \qquad (6.8)$$

To obtain equation (6.8) we introduced the following definitions

$$f_2(\lambda) = \frac{\varphi(\lambda)\tilde{t}_2(\lambda - \tfrac{1}{2})}{\psi(\lambda + \tfrac{1}{2})\psi(\lambda - \tfrac{1}{2})} \qquad (6.9)$$

$$b(\lambda) = \frac{\varphi(\lambda - 1)\varphi(\lambda + 1)}{\psi(\lambda + \tfrac{1}{2})\psi(\lambda - \tfrac{1}{2})}\frac{B(\lambda + \tfrac{1}{2})}{4}$$

$$= \left[\frac{(\lambda + s_\ell - \tfrac{1}{2})(\lambda - s_\ell + \tfrac{3}{2})}{(\lambda - s_\ell + \tfrac{1}{2})(\lambda + s_\ell + \tfrac{1}{2})}\right]\frac{B(\lambda + \tfrac{1}{2})}{4\lambda(\lambda + 1)(\lambda + \tfrac{1}{2})^2}. \qquad (6.10)$$

6.1. Single Non-linear Integral Equation

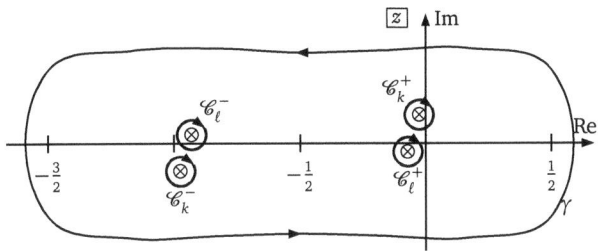

Figure 6.1: Integration contours $\Gamma_\ell^\pm = \mathscr{C}_\ell^\pm \cup \gamma$ and $\Gamma_k^\pm = \mathscr{C}_k^\pm \cup \gamma$. $\otimes$ denotes a singularity at s_ℓ or $-s_\ell - 1$ respectively.

Note that by this choice of $\psi(\lambda)$ we find $f_1(\lambda)$ to be a rational function with constant asymptotics i.e. it is expandable in partial fractions

$$f_1(z) = \frac{\operatorname{ch}(\phi)}{\alpha^+\alpha^-} + \sum_{\ell=0}^{L}\left[\frac{a_\ell^+}{z-s_\ell} + \frac{a_\ell^-}{z+s_\ell+1}\right], \quad a_\ell^\pm \in \mathbb{C}. \tag{6.11}$$

Here and from now on we use $s_0 \equiv 0$. Cauchy's integral formula can be applied to write these fractions as integrals. Each integration contour $\Gamma_\ell^\pm$ is a cycle homologous to zero which just excludes the respective singularity in $\mathbb{C}$. The argument z needs to be inside Γ. We then have

$$f_1(z) = \frac{\operatorname{ch}(\phi)}{\alpha^+\alpha^-} + \frac{1}{2\pi i}\sum_{\ell=0}^{L}\left[\int_{\Gamma_\ell^+}\frac{\mathrm{d}\xi}{(\xi-z)}\frac{a_\ell^+}{(\xi-s_\ell)} + \int_{\Gamma_\ell^-}\frac{\mathrm{d}\xi}{(\xi-z)}\frac{a_\ell^-}{(\xi+s_\ell+1)}\right]. \tag{6.12}$$

Introducing $\mathscr{C}$ as a small circle around the origin we can separate each $\Gamma_\ell^\pm$ into γ and a circle around the singularities $\mathscr{C}_\ell^+ \equiv \mathscr{C} + s_\ell$ or $\mathscr{C}_\ell^- \equiv \mathscr{C} - s_\ell - 1$ (see figure 6.1).

$$\begin{aligned}f_1(z) = {} & \frac{\operatorname{ch}(\phi)}{\alpha^+\alpha^-} + \frac{1}{2\pi i}\int_\gamma \frac{\mathrm{d}\xi}{(\xi-z)}\sum_{\ell=0}^{L}\left(\frac{a_\ell^+}{\xi-s_\ell} + \frac{a_\ell^-}{\xi+s_\ell+1}\right) \\ & - \frac{1}{2\pi i}\sum_{\ell=0}^{L}\left[\int_{\mathscr{C}_\ell^+}\frac{\mathrm{d}\xi}{(\xi-z)}\frac{a_\ell^+}{(\xi-s_\ell)} + \int_{\mathscr{C}_\ell^-}\frac{\mathrm{d}\xi}{(\xi-z)}\frac{a_\ell^-}{(\xi+s_\ell+1)}\right].\end{aligned} \tag{6.13}$$

By pushing γ to infinity or the use of residue theorem the integral over γ vanishes. As z is outside of each $\mathscr{C}_\ell^\pm$ we have

$$0 = \frac{1}{2\pi i}\int_{\mathscr{C}_\ell^\pm}\frac{\mathrm{d}\xi}{\xi-z}\frac{\operatorname{ch}(\phi)}{\alpha^+\alpha^-} + \frac{1}{2\pi i}\sum_{m\neq\ell}\int_{\mathscr{C}_\ell^\pm}\frac{\mathrm{d}\xi\, a_m^\pm}{(\xi-z)(\xi\pm s_\ell \pm \frac{1}{2} - \frac{1}{2})} \tag{6.14}$$

65

6. Non-linear Integral Equations for the XXX Spin Chain

and hence by adding such zeros each integrand of the remaining sum in (6.13) can be written as f_1 leading to

$$f_1(z) = \frac{\text{ch}(\phi)}{\alpha^+\alpha^-} + \frac{1}{2\pi i}\sum_{\ell=0}^{L}\left[\int_{\mathscr{C}_\ell^+}\frac{d\xi}{z-\xi}f_1(\xi) + \int_{\mathscr{C}_\ell^-}\frac{d\xi}{z-\xi}f_1(\xi)\right]$$

$$= \frac{\text{ch}(\phi)}{\alpha^+\alpha^-} + \frac{1}{2\pi i}\sum_{\ell=0}^{L}\left[\int_{\mathscr{C}+s_\ell-\frac{1}{2}}\frac{d\xi}{z-\xi-\frac{1}{2}}f_1(\xi+\tfrac{1}{2})\right.$$

$$\left. + \int_{\mathscr{C}-s_\ell-\frac{1}{2}}\frac{d\xi}{z-\xi+\frac{1}{2}}f_1(\xi-\tfrac{1}{2})\right]. \qquad (6.15)$$

To construct non-linear integral equations valid only for f_1 related to eigenvalues of the transfer matrix, we will analyze the singularities of $f_1(u+\tfrac{1}{2})$ and $f_1(u-\tfrac{1}{2})$ using (6.8).

Dividing (6.8) by $f_1(\lambda-\tfrac{1}{2})$

$$f_1(\lambda+\tfrac{1}{2}) = \frac{f_2(\lambda)}{f_1(\lambda-\tfrac{1}{2})} + \frac{b(\lambda)}{f_1(\lambda-\tfrac{1}{2})} \qquad (6.16)$$

leaves the left hand side with singularities at $\{s_\ell-\tfrac{1}{2}, -s_\ell-\tfrac{3}{2}, -\tfrac{1}{2}, 0\}$. On the right hand side we find

$$\frac{b(\lambda)}{f_1(\lambda-\tfrac{1}{2})} = \left[\prod_{\ell=1}^{L}\frac{(\lambda-s_\ell-\tfrac{1}{2})(\lambda+s_\ell-\tfrac{1}{2})(\lambda-s_\ell+\tfrac{3}{2})}{(\lambda-s_\ell+\tfrac{1}{2})}\right]\frac{(\lambda-\tfrac{1}{2})B(\lambda+\tfrac{1}{2})}{4(\lambda+\tfrac{1}{2})(\lambda+1)\widetilde{t}_1(\lambda-\tfrac{1}{2})} \qquad (6.17)$$

with singularities at $\{s_\ell-\tfrac{1}{2}, -\tfrac{1}{2}, -1, \text{zeros of } \widetilde{t}_1(\lambda-\tfrac{1}{2})\}$ and

$$\frac{f_2(\lambda)}{f_1(\lambda-\tfrac{1}{2})} = \left[\prod_{\ell=1}^{L}\frac{\lambda+s_\ell+\tfrac{1}{2}}{\lambda+s_\ell+\tfrac{3}{2}}\right]\frac{1}{(\lambda+1)(\lambda+\tfrac{3}{2})}\frac{\widetilde{t}_2(\lambda-\tfrac{1}{2})}{\widetilde{t}_1(\lambda-\tfrac{1}{2})} \qquad (6.18)$$

has singularities at $\{-s_\ell-\tfrac{3}{2}, -\tfrac{3}{2}, -1, \text{zeros of } \widetilde{t}_1(\lambda-\tfrac{1}{2})\}$. As the latter set does not include the arguments $\lambda = s_\ell-\tfrac{1}{2}$ and $\lambda = -\tfrac{1}{2}$ we put (6.16) into those integrals of (6.15) with the contours circling $\lambda = s_\ell-\tfrac{1}{2}$. Using the residue theorem the terms containing f_2 vanish and we are left with

$$f_1(z) = \frac{\text{ch}(\phi)}{\alpha^+\alpha^-} + \frac{1}{2\pi i}\sum_{\ell=0}^{L}\left[\int_{\mathscr{C}+s_\ell-\frac{1}{2}}\frac{d\xi}{z-\xi-\tfrac{1}{2}}\frac{b(\xi)}{f_1(z-\tfrac{1}{2})}\right.$$

$$\left. + \int_{\mathscr{C}-s_\ell-\frac{1}{2}}\frac{d\xi}{z-\xi+\tfrac{1}{2}}f_1(\xi-\tfrac{1}{2})\right]. \qquad (6.19)$$

Analogously dividing by the other factor we have

$$f_1(\lambda-\tfrac{1}{2}) = \frac{f_2(\lambda)}{f_1(\lambda+\tfrac{1}{2})} + \frac{b(\lambda)}{f_1(\lambda+\tfrac{1}{2})}. \qquad (6.20)$$

Examining the terms on the right hand side separately we have

$$\frac{b(\lambda)}{f_1(\lambda+\frac{1}{2})} = \left[\prod_{\ell=1}^{L}\frac{(\lambda+s_\ell-\frac{1}{2})(\lambda+s_\ell+\frac{3}{2})(\lambda-s_\ell+\frac{3}{2})}{(\lambda+s_\ell+\frac{1}{2})}\right]\frac{(\lambda+\frac{3}{2})B(\lambda+\frac{1}{2})}{4\lambda(\lambda+\frac{1}{2})\tilde{t}_1(\lambda+\frac{1}{2})} \quad (6.21)$$

with singularities at $\{-s_\ell-\frac{1}{2},-\frac{1}{2},0, \text{zeros of }\tilde{t}_1(\lambda-\frac{1}{2})\}$ and

$$\frac{f_2(\lambda)}{f_1(\lambda-\frac{1}{2})} = \left[\prod_{\ell=1}^{L}\frac{\lambda-s_\ell+\frac{1}{2}}{\lambda-s_\ell-\frac{1}{2}}\right]\frac{1}{\lambda(\lambda-\frac{1}{2})}\frac{\tilde{t}_2(\lambda-\frac{1}{2})}{\tilde{t}_1(\lambda+\frac{1}{2})} \quad (6.22)$$

with singularities at $\{s_\ell+\frac{1}{2},+\frac{1}{2},0, \text{zeros of }\tilde{t}_1(\lambda+\frac{1}{2})\}$. Replacing the integrand of the other integrals in (6.15), again the terms containing f_2 do not contribute because of the pole structure. We finally obtain

$$f_1(z) = \frac{\text{ch}(\phi)}{\alpha^+\alpha^-} + \frac{1}{2\pi i}\sum_{\ell=0}^{L}\left[\oint_{\mathscr{C}+s_\ell-\frac{1}{2}}\frac{d\xi}{z-\xi-\frac{1}{2}}\frac{b(\xi)}{f_1(\xi-\frac{1}{2})}\right.$$
$$\left.+\oint_{\mathscr{C}-s_\ell-\frac{1}{2}}\frac{d\xi}{z-\xi+\frac{1}{2}}\frac{b(\xi)}{f_1(\xi+\frac{1}{2})}\right]. \quad (6.23)$$

In conclusion the non-linear integral equation (with $s_0 \equiv 0$) valid for all states is

$$f_1(\lambda) = \frac{\text{ch}(\phi)}{\alpha^+\alpha^-} + \frac{1}{2\pi i}\sum_{\ell=0}^{L}\left[\oint_{\mathscr{C}+s_\ell-\frac{1}{2}}\frac{d\xi}{\lambda-\xi-\frac{1}{2}}\frac{b(\xi)}{f_1(\xi-\frac{1}{2})}\right.$$
$$\left.+\oint_{\mathscr{C}-s_\ell-\frac{1}{2}}\frac{d\xi}{\lambda-\xi+\frac{1}{2}}\frac{b(\xi)}{f_1(\xi+\frac{1}{2})}\right]. \quad (6.24)$$

This kind of derivation of integral equations was used by Takahashi and Klümper to show the equivalence of the thermodynamic Bethe ansatz and the quantum transfer matrix method [88, 89]. But in this form it has not yet been used to explicitly obtain the eigenvalue numerically.

6.2 Y-System and Non-linear Integral Equations

A different approach to non-linear integral equations utilizes the complete fusion hierarchy in auxiliary space to formulate an infinite set of non-linear integral equations referred to as thermodynamic Bethe ansatz like (TBA-like) equations [86, 87]. This concept was introduced by Klümper and Pearce [60, 61, 80] and has been generalized amongst others by Zhou and Pearce [103] for restricted $A_{n-1}^{(1)}$ fused lattice models and finally Zhou [101] adapted it for the $U_q[sl(2)]$ invariant six vertex model with open boundary conditions where the fusion hierarchy truncates at a finite level. In this section we analyse this method for the *XXX* model with an infinite hierarchy and provide a proof of principle for the applicability of the method in certain parameter regimes.

6. Non-linear Integral Equations for the XXX Spin Chain

6.2.1 Fusion hierarchy and T-system

The construction of the NLIE is based on a different but equivalent form of the fusion hierarchy (2.73) or (2.85) respectively [11]

$$t_k(\lambda + \tfrac{ic}{2})t_k(\lambda - \tfrac{ic}{2}) = \prod_{\ell=1}^{k} \delta(\lambda - \tfrac{ic}{2} + \ell ic) + t_{k-1}(\lambda + \tfrac{ic}{2})t_{k+1}(\lambda - \tfrac{ic}{2}). \qquad (6.25)$$

A proof of the equivalence is given in appendix D. By introducing a level dependent shift of the spectral parameter to the transfer matrix eigenvalue $T_k(\lambda + k\tfrac{ic}{2}) = t_{k-1}(\lambda)$ we obtain

$$T_k(\lambda - \tfrac{ic}{2})T_k(\lambda + \tfrac{ic}{2}) = \prod_{\ell=0}^{k-2} \delta(\lambda - (k-1)\tfrac{ic}{2} + \ell ic) + T_{k-1}(\lambda)T_{k+1}(\lambda) \qquad (6.26)$$

which is referred to as T-system [66] and references therein or TBA equations [101, 102] as well.

6.2.2 Y-system and Fourier transformation

The system of non-linear integral equations in this approach is derived for products of the transfer matrix eigenvalues denoted as Y-functions. This Y-system, also called universal form of the so-called TBA equations [66], follows directly from the T-system (6.26). Defining

$$Y_k(\lambda) \equiv \frac{T_{k-1}(\lambda)T_{k+1}(\lambda)}{\prod_{\ell=1}^{k-1} \delta(\lambda - k\tfrac{ic}{2} + (2\ell-1)\tfrac{ic}{2})}, \quad Y_0 \equiv -1, \quad Y_1 \equiv 0 \qquad (6.27)$$

and explicitely calculating $(1 + Y_{k+1})(1 + Y_{k-1})$ we find

$$Y_k(\lambda - \tfrac{ic}{2})Y_k(\lambda + \tfrac{ic}{2}) = (1 + Y_{k-1}(\lambda))(1 + Y_{k+1}(\lambda)). \qquad (6.28)$$

Note that a scaling of $t_1(\lambda)$ leaves all functions $Y_k(\lambda)$ invariant because due to (2.85) it results in a scaling of δ as well

$$t_1(\lambda) \to \frac{t_1(\lambda)}{f(\lambda)}, \quad \delta(\lambda) \to \frac{\delta(\lambda)}{f(\lambda)f(\lambda - ic)}, \quad t_k(\lambda - (k-1)ic) \to \frac{t_k(\lambda - (k-1)ic)}{\prod_{\ell=0}^{k-1} f(\lambda - \ell ic)}. \qquad (6.29)$$

This is important for the case of open boundary conditions as a function $f(\lambda) = (2\lambda + ic)^{-1}$ makes it possible to allocate the zeros of the Y-functions uniquely to zeros of the transfer matrices t_k and the poles of the Y-functions to the zeros of the functions δ given in (2.83). Further note, that all Y-functions have constant asymptotic by construction.

The Y-system is reformulated as a set of NLIE using standard Fourier transform techniques. We will use the following definition for the Fourier transform

$$\mathscr{F}(f)(k) = \int_{-\infty}^{\infty} dx\, e^{-ikx} f(x), \quad f(x) = \int_{-\infty}^{\infty} \frac{dk}{2\pi} e^{ikx} \mathscr{F}(f)(k). \qquad (6.30)$$

6.2. Y-System and Non-linear Integral Equations

In Fourier space a shift of the argument of any function becomes a prefactor. However, if the function is not analytical in the strip between the imaginary part of the shift and the real axis additive contributions arise. These contributions can be calculated by the residue theorem if the positions of the poles are known. As an example consider a shift $a \in \mathbb{R}$ and a Fourier transformable function f with simple poles $\{z_i\}$ and $\text{Im}(z_i) \in [0, \pm a]$ then (see appendix E for details)

$$\int_{-\infty}^{\infty} dx\, e^{-ikx} f(x \pm ia) = e^{\mp a}\mathcal{F}(f)(k) + 2\pi i \sum_i \text{Res}(e^{ikx} f(x), x = z_i). \tag{6.31}$$

To apply this to the left hand side of the Y-system (6.28) we need to apply a logarithmic derivative first,

$$\partial_\lambda \ln Y_k(\lambda + \tfrac{ic}{2}) + \partial_\lambda \ln Y_k(\lambda - \tfrac{ic}{2}) = \partial_\lambda \ln(1 + Y_{k-1}(\lambda)) + \partial_\lambda \ln(1 + Y_{k+1}(\lambda)). \tag{6.32}$$

The zeros and higher order poles (e.g. order n_i) turn into poles of first order and the applicability of Fourier transformations is assured leading to

$$2\text{ch}(kc/2)\mathcal{F}(\partial_\lambda \ln Y_k)(k) + 2\pi i \sum_i (\pm n_i) \text{Res}(e^{ik\lambda} \partial_\lambda \ln Y_k(\lambda), \lambda = z_i) =$$
$$\mathcal{F}(\partial_\lambda \ln(1 + Y_{k-1}))(k) + \mathcal{F}(\partial_\lambda \ln(1 + Y_{k+1}))(k) \tag{6.33}$$

Returning to the original space, the products of the inverse shift-factor and a function become convolution integrals with the kernel function

$$s(x) = \frac{1}{2\text{ch}(\pi x/c)}, \quad \mathcal{F}(s)(k) = \frac{1}{2\text{ch}(kc/2)} \tag{6.34}$$

while the contributions from poles and zeros are additive driving terms. These model-dependent driving terms are the challenging component of the problem.

Collecting all driving terms in the symbol $d_i(x)$ and integrating, the non-linear integral equations can be written as

$$\ln Y_2(x) = d_2(x) + (s * \ln(1 + Y_3))(x) \tag{6.35}$$
$$\ln Y_k(x) = d_k(x) + (s * \ln(1 + Y_{k-1}))(x) + (s * \ln(1 + Y_{k+1}))(x). \tag{6.36}$$

This system can be solved numerically by iteration if the driving terms are known and the behaviour of Y_k for large k is predictable.

The Y_2-function is then used to determine the eigenvalue of the transfer matrix generating the hamiltonian. With the same method described above the T-system at lowest level,

$$T_2(\lambda + ic)T_2(\lambda) = t_1(\lambda - ic)t_1(\lambda) = \delta(\lambda)(1 + Y_2(\lambda)), \tag{6.37}$$

is transformed to

$$\frac{\partial \ln t_1}{\partial \lambda}(\lambda - \tfrac{ic}{2}) = \int_{-\infty}^{\infty} \frac{dk}{2\pi} \frac{e^{ik\lambda}}{2\text{ch}(\tfrac{kc}{2})} \mathcal{F}(\partial_\lambda \ln \delta)(k) + e(\lambda) \tag{6.38}$$
$$+ (s * \partial \ln(1 + Y_2))(\lambda).$$

6. Non-linear Integral Equations for the XXX Spin Chain

Here t_1 is the eigenvalue of the transfer matrix and $e(\lambda)$ again denotes possible contributions due to poles and zeros of t_1 inside the integration contour used to draw the arguments on the left hand side together. In the following we will apply this program to spin chains with periodic and open boundaries.

6.2.3 NLIE for periodic boundaries

In the case of periodic boundary conditions with even chain length L it is easily numerically justified that for the ground state only the function $(\ln Y_2)$ has poles inside the strip $|\operatorname{Im}(\lambda)| \leq \frac{1}{2}$ located at $\lambda = 0$ with a multiplicity of L. The corresponding driving term d_2 is calculated straightforward in appendix E and all other $d_k = 0$ for $k > 2$. The equations read for even L

$$\ln Y_2(x) = L \ln|\tanh(\tfrac{\pi x}{2})| + (s * \ln(1 + Y_3))(x)$$
$$\ln Y_k(x) = (s * \ln(1 + Y_{k-1}))(x) + (s * \ln(1 + Y_{k+1}))(x) \,. \tag{6.39}$$

Notice that the constant asymptotics $Y_k \sim Y_k^\infty = k^2 - 1$ are a solution of (6.39) for $k > 2$ and can be used to approximate the system for large k.

In this case we have $\delta(\lambda) = (d_q T)(\lambda - ic) = (\lambda + ic)^L(\lambda - ic)^L$ from (2.73) and with $e(\lambda) \equiv 0$, c.f. (6.38), turns into

$$\frac{\partial \ln t_1}{\partial \lambda}(\lambda) = \frac{L}{ic}(1 - 2\ln 2) + \left(s * \partial \ln(1 + Y_2)\right)(\lambda + \tfrac{ic}{2}) \,. \tag{6.40}$$

Remembering that $\mathcal{H} = ic\, \partial_\lambda \ln t(0)$ we find by further manipulations of the convolution integral the ground state energy

$$E = L(1 - 2\ln 2) - \tfrac{1}{2}\int_{-\infty}^\infty dx\, \frac{\operatorname{ch} x}{\operatorname{sh}^2 x}\ln(1 + Y_2(\tfrac{xc}{\pi})) \tag{6.41}$$

recognizing its well known bulk behaviour.

6.2.4 NLIE for open boundaries

For open boundary conditions the situation is more subtle. First we will analyze the zero and pole structure of the Y-functions in the strip $|\operatorname{Im}(\lambda)| \leq \frac{1}{2}$ to identify the occurring driving terms. While the location of poles in the strip is clear for all states the distribution of zeros is predictable for $\phi \in \mathbb{R}$ defined in (2.86) and $\operatorname{sign}(\operatorname{Im}(\alpha^+)\operatorname{Im}(\alpha^-)) = -1$. We will restrict ourselves to this choice of parameters. Additionally we will set $c = 1$ without loss of generality for this section.

Using (6.29) to scale the transfer matrix properly only the denominators

$$Y_2 : \delta(\lambda), \quad Y_3 : \delta(\lambda + \tfrac{i}{2}), \quad Y_4 : \delta(\lambda - i)\delta(\lambda)\delta(\lambda + i), \quad \ldots \tag{6.42}$$

are responsible for poles of the Y-functions. Already from the form of the arguments of each δ given in (2.83) it is clear that the positions of poles vary by $i/2$ for successive k and

6.2. Y-System and Non-linear Integral Equations

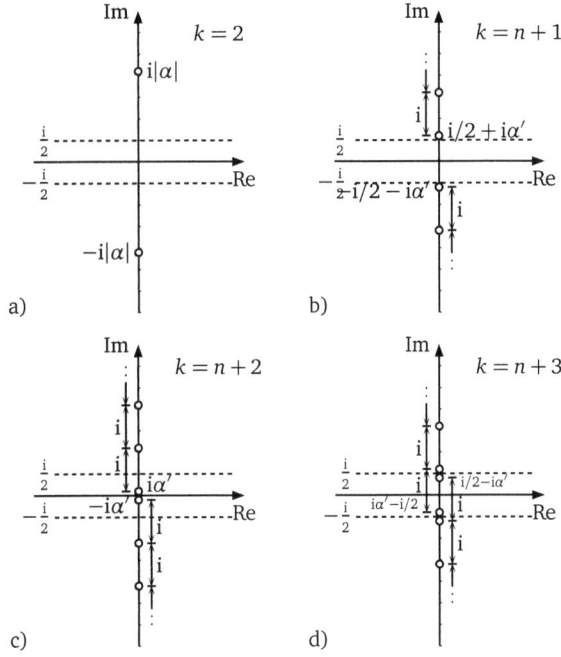

Figure 6.2: Exemplary position of poles (○) of Y_k in the complex plane arising from one boundary field α for a) $k = 2$, b) $k = n+1$, c) $k = n+2$ and d) $k = n+3$. With this examples all different cases of pole positions inside the strip $|\text{Im}(z)| \leq \frac{i}{2}$ are already considered.

two poles of the same k differ by i. An example is depicted in figure 6.2. Thus using the parametrization $|\alpha| = \frac{n}{2} + \alpha'$, $n \in \mathbb{N}, 0 < \alpha' < \frac{1}{2}$ the function $Y_k(\lambda)$ has poles at $\lambda = \pm i\alpha'$ for indices $k = n+2$ as a result. The according driving terms arising due to contracting the arguments on the left hand side are negative logarithms of (see appendix E)

$$h_c(x) \equiv \left(\frac{\text{ch}(\pi x) - \cos(\pi \alpha')}{\text{ch}(\pi x) + \cos(\pi \alpha')} \right) \quad \text{for } k = n+2, n+4, \ldots \quad (6.43)$$

$$h_s(x) \equiv \left(\frac{\text{ch}(\pi x) - \sin(\pi \alpha')}{\text{ch}(\pi x) + \sin(\pi \alpha')} \right) \quad \text{for } k = n+3, n+5, \ldots . \quad (6.44)$$

Note that only the modulus of the parameter α enters and the term h_c and h_s alternate in the equations of (6.46).

Turning to the zeros of the Y-functions we find that each but Y_2 has a double zero at $\lambda = 0$. Y_2 itself has a $2L+2$-fold zero at $\lambda = 0$. To be able to proceed we have to choose a particular state for which the transfer matrix does not have any further zeros in the strip of $|\text{Im} z| < \frac{1}{2}$. Then all Y-functions inherit this feature. This particular state exists for converse

6. Non-linear Integral Equations for the XXX Spin Chain

signs of the boundary parameters $\alpha^\pm$ and $\phi \in \mathbb{R}$ and in the limit of $\phi \to 0$ (i.e. diagonal boundary conditions) it becomes the state with lowest energy in the sector of vanishing magnetization.

A zero at $\lambda = 0$ gives a driving term of $\ln\tanh(|\frac{\pi x}{2}|)$ for even L and due to logarithmic derivative the multiplicity is only reflected in the integer prefactor. Bringing the common driving term $\ln\tanh^2(\frac{\pi x}{2})$ of all equations of (6.46) to the left hand side and arranging the driving terms in a matrix

$$b_k^n(x) = \begin{array}{c|cccccc} & 1 & 2 & 3 & 4 & \cdots & k \\ \hline 0 & \tanh^{2L}(\frac{\pi x}{2}) & h_c(x) & h_s(x) & h_c(x) & \cdots & \\ 1 & 1 & 1 & h_c(x) & h_s(x) & \cdots & \\ 2 & 1 & 1 & 1 & h_c(x) & \cdots & \\ \vdots & \vdots & \vdots & \vdots & \vdots & \ddots & \\ n & & & & & & \end{array} \qquad (6.45)$$

allows to write the non-linear integral equations in a compact form. For boundary fields $\alpha^+ = i|\alpha^+|$ and $\alpha^- = -i|\alpha^-|$ with $|\alpha^+| = \frac{n^+}{2} + \alpha'^+$ and $|\alpha^-| = \frac{n^-}{2} + \alpha'^-$ and $n^\pm \in \mathbb{N}$ we find for an asymptotic $\text{ch}\,\phi \geq 1$ of the transfer matrix the infinite set of equations

$$\ln\left(\frac{Y_2(x)}{\tanh^2(\frac{\pi x}{2})}\right) = \ln b_1^0(x) - \ln b_2^{n^+}(x) - \ln b_2^{n^-}(x) + (s*\ln(1+Y_3))(x)$$

$$\ln\left(\frac{Y_k(x)}{\tanh^2(\frac{\pi x}{2})}\right) = -\ln b_k^{n^+}(x) - \ln b_k^{n^-}(x) \qquad (6.46)$$

$$+ (s*\ln(1+Y_{k-1}))(x) + (s*\ln(1+Y_{k+1}))(x).$$

The asymptotics of the Y-functions $Y_k \sim Y_k^\infty = \frac{\text{sh}^2(k\phi)}{\text{sh}^2\phi} - 1$ solve the hierarchy (6.46) in the limit of $x \to \pm\infty$. In contrast to the periodic case the asymptotics are not a solution for arbitrary x of the equations for $k > 2$.

To obtain the eigenvalue of the hamiltonian we have to evaluate (6.38). For the choice of boundary parameters and particular state discussed there arise no additional pole contributions, $e(\lambda) \equiv 0$. The integral over the derivative of the Fourier transform of δ already arose in the context of separation of variables in (4.42). It was calculated by drawing together the arguments on the left hand side just as the Y-system was treated above (see appendix F for details) and explicitly is

$$i\frac{\partial \ln \Lambda_g}{\partial \lambda}\left(\frac{i}{2}\right) = (2-4\ln 2)L + \pi - 2\ln 2 - 1$$

$$+ \psi(\tfrac{|\alpha^+|}{2}) - \psi(\tfrac{|\alpha^+|}{2}+\tfrac{1}{2}) + \frac{1}{|\alpha^+|} \qquad (6.47)$$

$$+ \psi(\tfrac{|\alpha^-|}{2}) - \psi(\tfrac{|\alpha^-|}{2}+\tfrac{1}{2}) + \frac{1}{|\alpha^-|}.$$

Rolling the derivative in the convolution off to the kernel function $s(\lambda)$ and evaluating at $\lambda = 0$ leaves us with the eigenvalue of the hamiltonian (1.2)

$$i\frac{\partial \ln t_1}{\partial \lambda}(0) = i\frac{\partial \ln \Lambda_g}{\partial \lambda}\left(\frac{i}{2}\right) + \int_{-\infty}^{\infty} \frac{dk}{4\pi} k \tanh(\frac{k}{2}) \mathscr{F}(\ln(1+Y_2))(k) \tag{6.48}$$

$$= i\frac{\partial \ln \Lambda_g}{\partial \lambda}\left(\frac{i}{2}\right) + \int_{-\infty}^{\infty} dx \frac{\operatorname{ch}(\pi x)}{\operatorname{sh}^2(\pi x)} \ln(1+Y_2)(x). \tag{6.49}$$

6.2.5 Numerical situation

A system of infinite equations cannot be solved numerically without an approximation. In order to allow for numerical treatment we need an idea of how the solution Y_k behaves for large integers k. In the case of periodic boundary conditions it is clear that a good approximation is the constant asymptotic as it is a solution to the non-linear integral equation without driving term. Indeed with this constant limit function the iteration of (6.39) converges quickly and the accuracy can be controlled by numerically parameters (number of equations, size of position space, number of supporting points and accuracy to iterate the system to).

In the case of open boundary conditions the asymptotics are not a solution of any of the non-linear equations (6.46) for arbitrary arguments x as in every equation appears a driving term. Nevertheless using the constant asymptotic of Y_k^∞ as a limit function seems to assure convergence of the system well enough. Further we found that in the special cases of diagonal boundaries it is better to scale the constant asymptotic by the necessary driving terms in order to gain higher accuracy or use less equations without losing accuracy.

For asymptotic parameters $\phi \neq 0$ the asymptotic

$$Y_k^\infty = \frac{\operatorname{sh}((k+1)\phi)\operatorname{sh}((k-1)\phi)}{\operatorname{sh}^2 \phi} \tag{6.50}$$

grows exponentially with k and numerical limitations are very quickly reached. The number of non-linear integral equations, necessary for decent accuracy, produces function values for which more sophisticated numerical treatment is needed.

A more comprehensible approach than just using the asymptotic as a limit function is to look for solutions of (6.46) in the limit of $k \to \infty$. As the system of equations (6.46) has alternating driving terms we define two limit functions $g_1(x)$ and $g_2(x)$ by

$$\begin{aligned} Y_k(x) &= Y_k^\infty \frac{\tanh^2(\frac{\pi x}{2}) g_1(x)}{b_k^{n^+}(x) b_k^{n^-}(x)} \\ Y_{k'}(x) &= Y_{k'}^\infty \frac{\tanh^2(\frac{\pi x}{2}) g_2(x)}{b_{k'}^{n^+}(x) b_{k'}^{n^-}(x)} \end{aligned} \tag{6.51}$$

6. Non-linear Integral Equations for the XXX Spin Chain

for successive k and $k' = k+1$. Inserting these into (6.46) we obtain two coupled non-linear integral equations

$$\ln g_1 = s * \ln \left(\frac{1 + Y^\infty_{k-1} \frac{\tanh^2(\frac{\pi\circ}{2}) g_2}{b^{n+}_{k-1} b^{n-}_{k-1}}}{1 + Y^\infty_{k-1}} \cdot \frac{1 + Y^\infty_{k+1} \frac{\tanh^2(\frac{\pi\circ}{2}) g_2(x)}{b^{n+}_{k+1} b^{n-}_{k+1}}}{1 + Y^\infty_{k+1}} \right)$$

$$\ln g_2 = s * \ln \left(\frac{1 + Y^\infty_{k'-1} \frac{\tanh^2(\frac{\pi\circ}{2}) g_1}{b^{n+}_{k'-1} b^{n-}_{k'-1}}}{1 + Y^\infty_{k'-1}} \cdot \frac{1 + Y^\infty_{k'+1} \frac{\tanh^2(\frac{\pi\circ}{2}) g_2}{b^{n+}_{k'+1} b^{n-}_{k'+1}}}{1 + Y^\infty_{k'+1}} \right) .$$

(6.52)

In the limit of $k \to \infty$ the system linearizes

$$\ln g_1 = 2s * \ln g_2 + 2s * \ln \tanh^2(\tfrac{\pi\circ}{2}) - 2s * \ln b^{n+}_{k-1} - 2s * \ln b^{n-}_{k-1}$$
$$\ln g_2 = 2s * \ln g_1 + 2s * \ln \tanh^2(\tfrac{\pi\circ}{2}) - 2s * \ln b^{n+}_{k'-1} - 2s * \ln b^{n-}_{k'-1}$$

(6.53)

and is solvable by Fourier transformation yielding

$$g_1(x) = \left(2 \operatorname{ch}(\pi) \frac{\operatorname{ch}(\frac{\pi x}{2}) \operatorname{ch}(\frac{\pi x}{2})}{d^{n+}_k(x) d^{n-}_k(x)} \right)$$
$$g_2(x) = \left(2 \operatorname{ch}(\pi) \frac{\operatorname{ch}(\frac{\pi x}{2}) \operatorname{ch}(\frac{\pi x}{2})}{d^{n+}_{k+1}(x) d^{n-}_{k+1}(x)} \right)$$

(6.54)

with

$$d^n_k(x) = \begin{cases} \operatorname{ch}(\pi x) + \cos(\pi \alpha') & \text{for } n+k \text{ even} \\ \operatorname{ch}(\pi x) + \sin(\pi \alpha') & \text{for } n+k \text{ odd} \end{cases} .$$

(6.55)

Unfortunately this truncation does not allow to reduce the number of equations substantially and the numerical problems prevail.

6.2.6 Remarks

The parameter range $\phi \in i\mathbb{R}$ of hermitian hamiltonians is not reachable with the method described here as there arise further zeros in the strip $|\operatorname{Im}(z)| \leq \frac{1}{2}$ for the ground state. Furthermore some Y-functions even have function values below -1 and above $+1$ hence the non-linear integral equations (6.46) need to be treated as complex equations.

The fact that the solution presented here is only valid for $\operatorname{ch} \phi \geq 1$ coincides with a different approach starting from the fusion hierarchy which results in pseudo Bethe equations to be solved for the eigenvalue [38]. The approach is based on the fact that the fused transfer matrix eigenvalue will turn into a Q-function of the TQ-equation of the model [99] and makes use of a systematic predictability of zero distributions of the fused eigenvalues. In their approach the zero distributions in the parameter regime of ch $\phi < 1$ could not be dealt with.

In the case of the *XXZ* model with diagonal boundaries and a very special choice of parameters the fusion hierarchy truncates and the problem was already solved in [101]. But a solution for the general case and non-diagonal boundaries is an open problem just as the region $\phi \in i\mathbb{R}$ for the *XXX* model.

Chapter 7
Conclusion and Outlook

In this thesis we considered the spin-$\frac{1}{2}$ *XXZ* and *XXX* spin chain subject to boundary fields. For the Bethe ansatz solvable case of S^z-conserving diagonal boundary fields we derived a Bethe roots independent non-linear integral equation for an auxiliary function a accounting for the ground state of the model or a neighboring state of zero magnetization respectively of parameter choices. These equation is valid for a finite number of even lattice sites and was combined with the scalar product formula of Bethe vectors of Kitanine *et al.* [54] to address correlation functions.

This yielded a linear integral equation whose solution builds on the well known determinant representing scalar products. As an example we derived for a certain generating function of the S^z-magnetization a multiple integral representation showing the correct thermodynamic limit and matching the result from exact diagonalization for small lattice sites. For the derivation we included states with distributions of Bethe solutions having one hole-type solution on the real line. For this cases we had to choose a closed contour $\mathscr{C}'$ for the integral representations of the determinant formula and generating function differing from the canonical contour $\mathscr{C}$ of the auxiliary function. But this is already the general case. In the simpler issue of field parameters no 'holes' have to be taken into account such that the canonical contour even applies for the integral representations. Therefore we expect this to be a good starting point for numerical considerations.

The situation for non-diagonal boundary fields is not as advanced and we approached the spectral problem with two different approaches.

First we have extended Sklyanin's functional Bethe ansatz method to the open *XXX* chain with non-diagonal boundary fields. Within this framework we have derived a single *TQ*-equation (4.38) which determines the spectrum of this model for any values of the boundary parameters. The TQ-equation allows for a solution in terms of polynomials for the function Q provided that a constraint between the left and right boundary field is satisfied. In this case the solution is parametrized by the roots of one set of Bethe ansatz equations. If the constraint between the boundary fields is missing only the asymptotic (exponential) behaviour of the Q-functions is obtained from the TQ-equation while the sub-leading terms have to be chosen such that the eigenvalues of the transfer matrix remain polynomial. In Sklyanin's approach the Q-functions contain all the information on the eigenstates of the model. Therefore, their determination for generic boundary parameters is necessary to tackle the problem of computing norms and scalar products within this approach. However, unlike the situation with the algebraic Bethe ansatz, where one has an expression for the eigenstates in terms of the generators of the Yang-Baxter algebra, the explicit transformation from the Q-functions to state vectors in the Hilbert space of the spin chain is not known. To make progress in this direction it should be useful to investigate how the recent construction of Galleas [35] connects to the TQ-equation (4.38). His solution of the spectral problem for XXZ chains with non-diagonal boundaries is given in terms of the zeros of the transfer matrix eigenvalues and two complementary sets of numbers which parametrize matrix elements of certain entries of the monodromy matrix. They satisfy equations reminiscent of the nested Bethe ansatz used to solve models of higher-rank symmetries. Further studies are necessary to see whether this parametrization of the spectrum in terms of $\mathcal{O}(L)$ complex numbers can be used to obtain a closed expression for generic (non-polynomial) Q-functions. This would be of great importance for the applicability of the TQ-equation to solve the spectral problem of integrable quantum chains.

The other approach made use of the fusion hierarchy relating transfer matrices with different dimensional auxiliary spaces. From this functional relation two types of non-linear integral equations were derived. The first type only involves the first level of the hierarchy. The result is a single equation valid for all states of the model. It relates the scaled eigenvalue of the transfer matrix to its reciprocal at a shifted argument but it does not offer an obvious numerical treatment. The other uses the full infinite fusion hierarchy to derive a Y-system similar to the thermodynamic Bethe ansatz. For a particular state and parameter range the zero and pole structure of the Y-functions is predictable. In this case the Y-system is turned into an infinite set of non-linear integral equations by Fourier transformation. The valid parameter range renders the hamiltonian non-hermitian and the state described is the lowest lying state of zero magnetization in the limit to diagonal boundaries. With the asymptotics of the Y-functions an approximate truncation of infinite set of integral equations seems possible and a numerical treatment is within reach.

Besides the open spin chains we extended the approach of separation of variables to the spin-boson model in a straightforward fashion. For this model we were able to derive a TQ-equation valid on an infinite set of discrete points prohibiting numerical studies analog to the spin chain case. Interestingly the asymptotical analysis revealed a Γ-function dependence of the involved Q-functions which disappears in the Bethe ansatz solvable cases. Progress on the TQ-equation for the spin-boson model may be achieved by employing certain quasi-classical limits[1] turning it into a differential equation or further study of the half-infinite determinants arising when eliminating the Q-functions.

We have revisited the *XXZ* spin chain with anti-diagonal twist which does not allow for a solution of the spectral problem by means of the algebraic Bethe ansatz due to the lack of a reference state. We have derived the functional equations (5.6), originally obtained using Baxter's method of commuting transfer matrices, by truncating the fusion hierarchy and restricting the anisotropy to roots of unity $\eta = i\pi/(p+1)$. This setting allowed to compare Sklyanin's separation of variables method to the performed fusion hierarchy truncation approach first applied to spin chains by Nepomechie. Within both techniques the same TQ-equation was found. This is unlike the situation for the spin chain with open boundaries and non-diagonal boundary fields where different functional equations have been found within the different approaches revealing this as a model dependent feature.

---

[1] Unfortunately a quasi-classical limit suggested in [5] does not combine rotating and counter-rotating terms in a Hamiltonian as believed.

Appendices

Appendix A

Density Function for Diagonal Boundaries

We consider the set $\{\mu_\ell\}_{\ell=1}^M = \{v_j\}_{j=1}^n \cup \{\lambda_\ell\}_{\ell=n+1}^M$ in order to not overload the notation. Then the expression under the determinant of the right hand side of (3.19) reads in components

$$\psi(\lambda_j, v_\ell) = \frac{i}{\text{sh}(2\lambda_j)} \sum_{k=1}^M \left[K_\eta(\lambda_j + \lambda_k) - K_\eta(\lambda_j - \lambda_k) \right] J(\lambda_k, v_\ell) - \frac{J(\lambda_j, v_\ell)}{\text{sh}(2\lambda_j)} \frac{\partial \ln \mathfrak{a}}{\partial z}(\lambda_j), \quad (A.1)$$

solved for the part containing the logarithmic derivative of the auxiliary function

$$\begin{aligned}
J(\lambda_j, v_\ell) \frac{\partial \ln \mathfrak{a}}{\partial z}(\lambda_j) = & \; i \sum_{k=1}^M \left[K_\eta(\lambda_j + \lambda_k) - K_\eta(\lambda_j - \lambda_k) \right] J(\lambda_k, v_\ell) \\
& + \left[\frac{\text{sh}\,\eta}{\text{sh}(v_\ell - \lambda_j)\text{sh}(v_\ell - \lambda_j - \eta)} - \frac{\text{sh}\,\eta}{\text{sh}(v_\ell + \lambda_j)\text{sh}(v_\ell + \lambda_j - \eta)} \right] \\
& - \left[\frac{\text{sh}\,\eta}{\text{sh}(v_\ell - \lambda_j)\text{sh}(v_\ell - \lambda_j + \eta)} - \frac{\text{sh}\,\eta}{\text{sh}(v_\ell + \lambda_j)\text{sh}(v_\ell + \lambda_j + \eta)} \right] \mathfrak{a}(v_\ell). \quad (A.2)
\end{aligned}$$

Strikingly the definition of $J(\lambda, v)$ according to (A.2) is compatible with the constraint $J(\lambda, v) = -J(-\lambda, v)$ and the properties $\mathfrak{a}(z) = 1/\mathfrak{a}(-z)$, $\mathfrak{a}'(\lambda_j) = \mathfrak{a}'(-\lambda_j)$ of the auxiliary function. Considering

$$F(\lambda_j, v_\ell) \equiv J(\lambda_j, v_\ell) \frac{\partial \ln \mathfrak{a}}{\partial z}(\lambda_j) \quad (A.3)$$

A. Density Function for Diagonal Boundaries

for arbitrary arguments λ_j the analytic properties $\text{Res}_{\lambda=\pm\nu} F(\lambda,\nu) = 1 + \mathfrak{a}(\nu)$ beside the single zero $F(0,\nu) = 0$ are known from the right hand side of (A.2) such that

$$\int_{\mathscr{C}'} \frac{d\omega}{2\pi i} \frac{\text{sh}(2\eta)}{\text{sh}(\lambda-\omega+\eta)\text{sh}(\lambda-\omega-\eta)} \frac{F(\omega,\nu)}{1+\mathfrak{a}(\omega)} = F(\lambda,\nu)$$

$$+ (1+\mathfrak{a}(\nu))\left[\frac{\text{sh}\,\eta}{\text{sh}(\lambda+\nu)\text{sh}(\lambda+\nu-\eta)} + \frac{\text{sh}\,\eta}{\text{sh}(\lambda-\nu)\text{sh}(\lambda-\nu-\eta)}\right] \quad (A.4)$$

holds where all Bethe roots λ_j and the variable ν except the hole-type solution χ on the real line are supposed to lie inside $\mathscr{C}'$. Redefining $F(\lambda,\nu) \equiv -(1+\mathfrak{a}(\nu))G(\lambda,\nu)$ we are led to

$$G(\lambda,\nu) = \frac{\text{sh}\,\eta}{\text{sh}(\lambda-\nu)\text{sh}(\lambda-\nu-\eta)} + \frac{\text{sh}\,\eta}{\text{sh}(\lambda+\nu)\text{sh}(\lambda+\nu-\eta)}$$

$$+ \int_{\mathscr{C}'} \frac{d\omega}{2\pi i} \frac{\text{sh}(2\eta)}{\text{sh}(\lambda-\omega+\eta)\text{sh}(\lambda-\omega-\eta)} \frac{G(\omega,\nu)}{1+\mathfrak{a}(\omega)}. \quad (A.5)$$

Unfortunately in our case ν should be a lattice inhomogeneity ζ_j taken from the strip $|\text{Im}(\nu-\eta/2)| < \varepsilon$. Thus $G(\lambda,\nu)$ is due to an additional residue and $\mathfrak{a}(\zeta_j) = 0$ the solution to the linear integral equation (3.23). The considered determinant (3.19) is calculated from

$$\frac{\det\left[\psi(\lambda_a,\mu_b)\right]_{a,b=1,\ldots,M}}{\det\left[\phi(\lambda_j,\lambda_k)\right]_{j,k=1,\ldots,M}} = \det\left[J(\lambda_j,\nu_\ell)\right]_{j,\ell=1,\ldots,n}$$

$$= \det\left[\frac{(1+\mathfrak{a}(\nu_\ell))G(\lambda_j,\nu_\ell)}{\mathfrak{a}'(\lambda_j)}\right]_{j,\ell=1,\ldots,n} \quad (A.6)$$

and making again use of $\mathfrak{a}(\zeta_j) = 0$ for the variable $\nu_l = \zeta_j = \eta/2 + s_j$ reduces the expression (A.6) to (3.22). Clearly, explicitly using $\mathfrak{a}(\zeta_j) = 0$ changes $G(\lambda,\nu)$ for the argument ν away from ζ_j compared to the original definition. But because $G(\lambda,\nu)$ is only used in combination with simple poles and the residue theorem (c.f. (3.35)) just its unchanged value $G(\lambda,\zeta_j)$ at the lattice inhomogeneity ζ_j is relevant.

Appendix B

Comparison to Numerics for Diagonal Boundaries

The energy values of the Hamiltonian (1.1) for diagonal boundary fields, i.e. $\kappa^\pm = 0$, were obtained by iteration in [97]. We show data for solutions of system size $L = 42$, hence the Bethe ansatz equations (2.58) were solved in the sector $M = L/2 = 21$ or, for the boundary bound states, modified Bethe ansatz equations were solved with one or two fixed imaginary Bethe numbers and accordingly many real Bethe numbers. The anisotropy is fixed to $\gamma = \frac{i\pi}{3}$.

The figures B.1 to B.4 show on the left side the integration contour where the vertical axis is assigned to the position of the pole associated to the parameter ξ^+ for the interval $(-i\pi/2, i\pi/2]$. On the right side the numerical dated is presented where the axis and orientation is chosen such that both vertical axes correspond to each other. The figures point out the various regions listed in table 3.1 by means of the integration contour in the complex plane and the energy spectrum in arbitrary units. The boundary parameter ξ^- remains fixed for each figure while ξ^+ is the independent parameter. In the integration contour the pole arising from ξ^+ is located on the imaginary axis at $\frac{i\gamma}{2} - \xi^+$, i.e. the pole position is in the interval $(-i\pi/3, 2i\pi/3]$. The corresponding energy of the lowest lying state of zero magnetization may then be read off on the right panel from the lowest set of data points. One can clearly make out the arising of boundary bound states in the numerical data when the pole corresponding to ξ^+ reaches the real axis from below in the displayed sector of the integration contour. As described in chapter 3 and [97] the pole picks up a Bethe root and drags it along the imaginary axis. Reaching the top of the contour, this imaginary root disappears and a transition to a different description of the state takes place. Notice that the value of the boundary field at the top of the contour is $\xi^+ = 0$ which leads to an infinite prefactor

B. Comparison to Numerics for Diagonal Boundaries

of the operator σ_L^z in the boundary part of the hamiltonian. This singularity explains the behaviour of the numerical data approaching $\xi^+ = 0$. In conclusion the number of real and imaginary Bethe roots in the various sectors and the change of description of the state is numerically verified at this example.

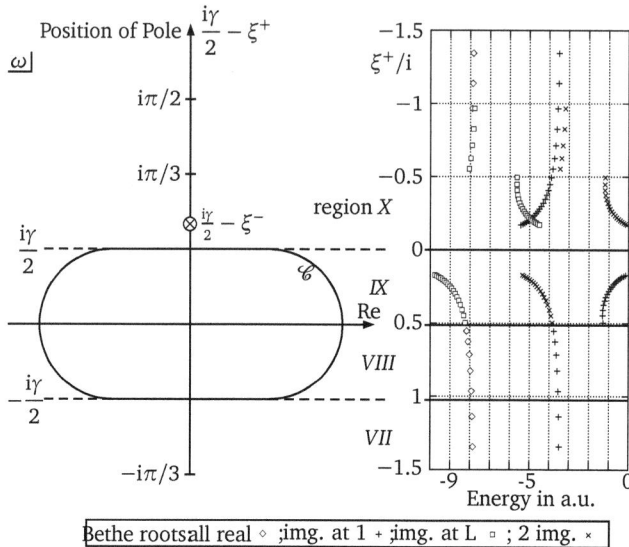

Figure B.1: Integration contour and energy spectrum for arbitrary ξ^+ with $L = 42$, $\xi^- = -0.213$ and $\eta = i\gamma = i\pi/3$. The numerical data is shown with different markers for different sectors of Bethe roots varying from 21 real roots to two imaginary roots and 19 real roots. The fixed pole arising from ξ^- is above the contour. Hence by varying ξ^+ its corresponding pole travels through the contour and the regions $VII, VIII, IX, X$ from table 3.1 are depicted. Note the changing of Bethe roots when the pole reaches the real axis from below.

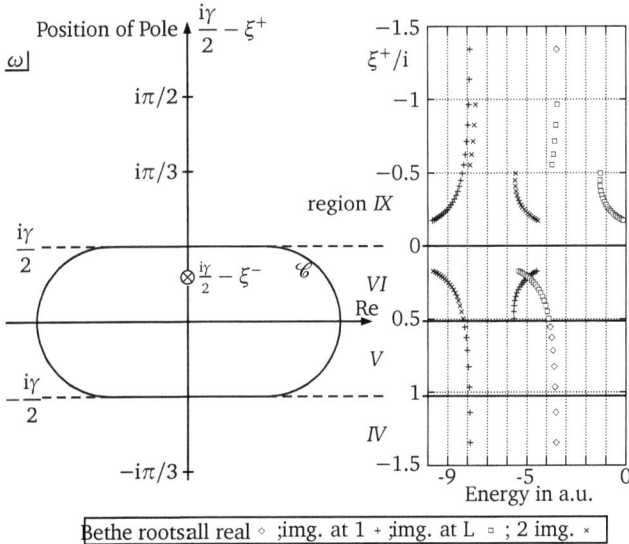

Figure B.2: Description see fig. B.1. Here $\xi^- = 0.213$ displaying regions IV, V, VI, IX.

B. COMPARISON TO NUMERICS FOR DIAGONAL BOUNDARIES

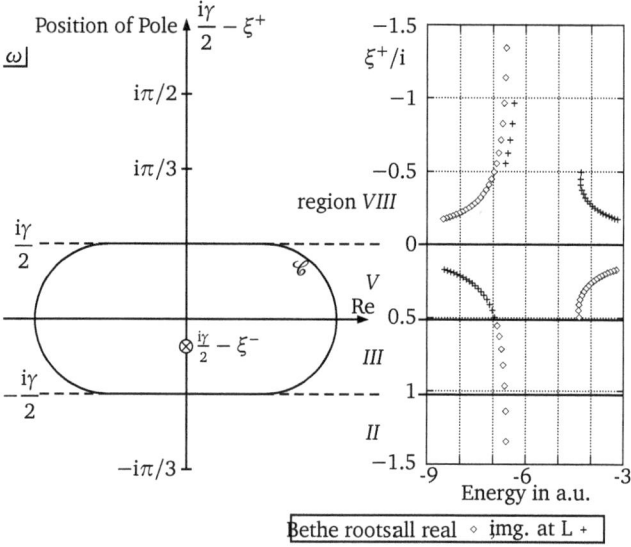

Figure B.3: Description see fig. B.1. Here $\xi^- = 0.713$ displaying regions $II, III, V, VIII$.

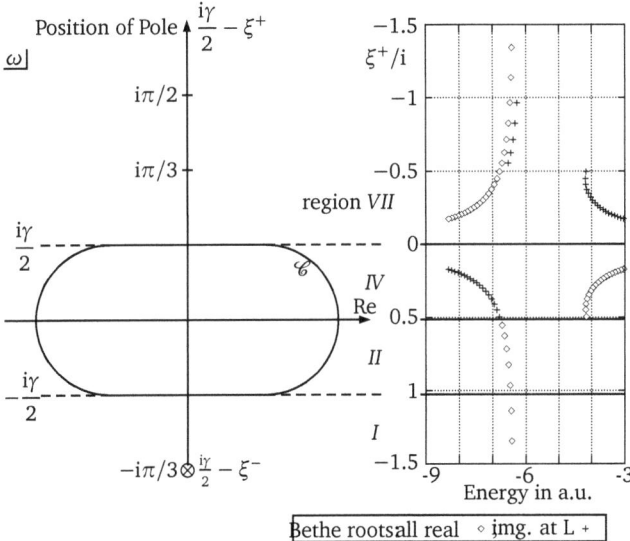

Figure B.4: Description see fig. B.1. Here $\xi^- = 1.571$ displaying regions I, II, IV, VII.

Appendix C

Similarity Transformations for Truncation Identity at Roots of Unity

In chapter 5 certain similarity transformations were used to relate the fused twist and R-matrices. As an example we want to give explicit expression for these transformation matrices for the first two cases.

For $k = 2$ we have

$$A = \begin{pmatrix} 1 & 0 & 0 & 0 \\ 0 & \frac{1}{2} & \frac{1}{2} & 0 \\ 0 & 0 & 0 & 1 \\ 0 & \frac{1}{2} & -\frac{1}{2} & 0 \end{pmatrix} \quad , \quad B = \begin{pmatrix} 1 & 0 & 0 & 0 \\ 0 & a & 0 & 0 \\ 0 & 0 & a & 0 \\ 0 & 0 & 0 & 1 \end{pmatrix} \tag{C.1}$$

with $a = ([2]_q)^{-\frac{1}{2}}$ where $q \equiv e^\eta$ is a root of unity, satisfying $q^{p+1} = -1$.

For $k = 3$ we have now with $a = ([3]_q)^{-\frac{1}{2}}$

$$A = \begin{pmatrix} 1 & 0 & 0 & 0 & 0 & 0 & 0 & 0 \\ 0 & \frac{1}{3} & \frac{1}{3} & 0 & \frac{1}{3} & 0 & 0 & 0 \\ 0 & 0 & 0 & \frac{1}{3} & 0 & \frac{1}{3} & \frac{1}{3} & 0 \\ 0 & 0 & 0 & 0 & 0 & 0 & 0 & 1 \\ 0 & -\frac{2}{3} & \frac{1}{3} & 0 & \frac{1}{3} & 0 & 0 & 0 \\ 0 & 0 & 0 & \frac{1}{3} & 0 & \frac{1}{3} & -\frac{2}{3} & 0 \\ 0 & 0 & 0 & \frac{1}{2} & 0-\frac{1}{2} & 0 & 0 & 0 \\ 0 & 0 & 0 & \frac{1}{2} & 0-\frac{1}{2} & 0 & 0 & 0 \end{pmatrix} \quad , \quad B = \mathrm{diag}(a, 1, 1, a, 1, 1, 1, 1). \tag{C.2}$$

Appendix D

Equivalent Form of Fusion Hierarchy

Proposition.

$$t_k(\lambda + \tfrac{\eta}{2})t_k(\lambda - \tfrac{\eta}{2}) = \prod_{\ell=1}^{k} \delta(\lambda - \tfrac{\eta}{2} + \ell\eta) + t_{k-1}(\lambda + \tfrac{\eta}{2})t_{k+1}(\lambda - \tfrac{\eta}{2}) \tag{D.1}$$

is equivalent to the fusion hierarchy

$$t_k(\lambda) = t_{k-1}(\lambda)t_1(\lambda + (k-1)\eta) - \delta(\lambda + (k-1)\eta)t_{k-2}(\lambda), \quad k = 2, 3, \ldots \tag{D.2}$$

with $t_0(\lambda) = 1, t_{-1} = 0$ and $t_1(\lambda) \equiv t(\lambda)$ being the eigenvalue of the regular transfer matrix.

Proof. The proof is carried out using mathematical induction.

Looking at (D.2) for $k = 2$

$$t_2(\lambda) = t_1(\lambda)t_1(\lambda + \eta) - \delta(\lambda + \eta)t_0(\lambda) \tag{D.3}$$

and using $t_0 \equiv 1$ yields the basis ($k = 1$) of (D.1) directly

$$t_1(\lambda - \tfrac{\eta}{2})t_1(\lambda + \tfrac{\eta}{2}) = \delta(\lambda + \tfrac{\eta}{2}) + t_0(\lambda + \tfrac{\eta}{2})t_2(\lambda - \tfrac{\eta}{2}). \tag{D.4}$$

The inductive hypothesis is given by (D.1).

D. Equivalent Form of Fusion Hierarchy

The inductive step starts with substituting t_{k+2} on the left hand by using (D.2)

$$
\begin{aligned}
t_k(\lambda+\tfrac{\eta}{2})t_{k+2}(\lambda-\tfrac{\eta}{2}) &= t_k(\lambda+\tfrac{\eta}{2})\Big[t_1(\lambda+k+\tfrac{\eta}{2})t_{k+1}(\lambda-\tfrac{\eta}{2}) \\
&\qquad - \delta(\lambda+k+\tfrac{\eta}{2})t_k(\lambda-\tfrac{\eta}{2})\Big] \\
&= \underbrace{t_k(\lambda+\tfrac{\eta}{2})t_1(\lambda+k+\tfrac{\eta}{2})}_{\text{using (D.2)}} t_{k+1}(\lambda-\tfrac{\eta}{2}) \\
&\quad - \delta(\lambda+k+\tfrac{\eta}{2})\underbrace{t_k(\lambda+\tfrac{\eta}{2})t_k(\lambda-\tfrac{\eta}{2})}_{\text{using hypothesis}} \\
&= t_{k+1}(\lambda+\tfrac{\eta}{2})t_{k+1}(\lambda-\tfrac{\eta}{2}) \\
&\quad + \delta(\lambda+k+\tfrac{\eta}{2})t_{k-1}(\lambda+\tfrac{\eta}{2})t_{k+1}(\lambda-\tfrac{\eta}{2}) \\
&\quad - \delta(\lambda+k+\tfrac{\eta}{2})t_{k-1}(\lambda+\tfrac{\eta}{2})t_{k+1}(\lambda-\tfrac{\eta}{2}) \\
&\quad - \delta(\lambda+k+\tfrac{\eta}{2})\prod_{\ell=1}^{k}\delta(\lambda-\tfrac{\eta}{2}+\ell) \\
&= t_{k+1}(\lambda+\tfrac{\eta}{2})t_{k+1}(\lambda-\tfrac{\eta}{2}) - \prod_{\ell=1}^{k+1}\delta(\lambda-\tfrac{\eta}{2}+\ell)
\end{aligned}
$$

which proves the assertion. $\square$

Appendix E

Calculation of Driving Terms

In this section we describe how to solve an equation of the form

$$Y(x+\tfrac{i}{2})Y(x-\tfrac{i}{2}) = \ldots \tag{E.1}$$

for $Y(x)$ where the right hand side is left unspecified on purpose and we have an $\chi \in \mathbb{C}$ with $y(\pm\chi) = 0$. The terms arising because of the zero χ are driving terms in the obtained integral equations. We consider the contours $\mathscr{C}_1$ and $\mathscr{C}_2$ in figure E.1 and use complex analysis to express the Fourier transform $\mathscr{F}(\partial \ln Y)(k) \equiv \int_{-\infty}^{\infty} dx\, e^{-ikx} \partial \ln Y(x)$. With the

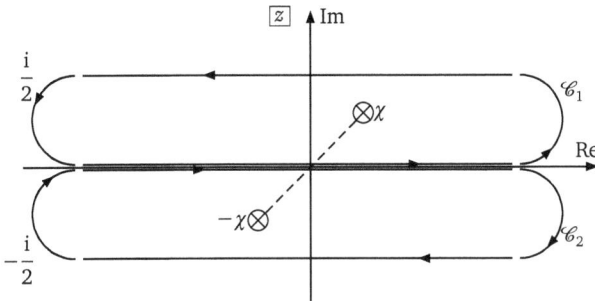

Figure E.1: Contours for solving the functional equation

E. Calculation of Driving Terms

residue theorem we find

$$\int_{\mathscr{C}_1} dz\, e^{-ikz} \partial \ln Y(z) = \int_{-\infty}^{\infty} dx\, e^{-ikx} \partial \ln Y(x) - \int_{-\infty}^{\infty} dx\, e^{-ik(x+\frac{i}{2})} \partial \ln Y(x+\tfrac{i}{2})$$

$$= \mathscr{F}(\partial \ln Y)(k) - e^{\frac{k}{2}} \int_{-\infty}^{\infty} dx\, e^{-ikx} \partial \ln Y(x+\tfrac{i}{2})$$

$$\stackrel{!}{=} +2\pi i e^{-ik\chi}$$

$$\int_{\mathscr{C}_2} dz\, e^{-ikz} \partial \ln Y(z) = \int_{-\infty}^{\infty} dx\, e^{-ikx} \partial \ln Y(x) - \int_{-\infty}^{\infty} dx\, e^{-ik(x-\frac{i}{2})} \partial \ln Y(x+\tfrac{i}{2})$$

$$= \mathscr{F}(\partial \ln Y)(k) - e^{-\frac{k}{2}} \int_{-\infty}^{\infty} dx\, e^{-ikx} \partial \ln Y(x+\tfrac{i}{2})$$

$$\stackrel{!}{=} -2\pi i e^{-ik\chi}$$

resulting in

$$\int_{-\infty}^{\infty} e^{-ikx} \partial \ln Y(x+\tfrac{i}{2}) = e^{-\frac{k}{2}} \mathscr{F}(\partial \ln Y)(k) - 2\pi i e^{-\frac{k}{2}} e^{-ik\chi}, \quad (E.2)$$

$$\int_{-\infty}^{\infty} e^{-ikx} \partial \ln Y(x-\tfrac{i}{2}) = e^{+\frac{k}{2}} \mathscr{F}(\partial \ln Y)(k) + 2\pi i e^{+\frac{k}{2}} e^{-ik\chi}. \quad (E.3)$$

Adding the two equations and solving for the wanted Fourier transform we find

$$\mathscr{F}(\partial \ln Y)(k) = -2\pi i \frac{\operatorname{sh}(\frac{k}{2}+ik\chi)}{\operatorname{ch}(\frac{k}{2})} + \frac{1}{2\operatorname{ch}(\frac{k}{2})} \int_{-\infty}^{\infty} dx\, e^{-ikx} \partial \ln\left(Y(x+\tfrac{i}{2})Y(x-\tfrac{i}{2})\right) \quad (E.4)$$

here the integral depends on the unspecified right hand side of (E.1). Introducing $v \equiv 1 - \frac{2\chi}{i}$ and turning to the Fourier transformation to position space for the known term we have

$$\int_{-\infty}^{\infty} \frac{dk}{2\pi} e^{ikx} \frac{\operatorname{sh}(\frac{vk}{2})}{\operatorname{ch}(\frac{k}{2})} = i \int_{-\infty}^{\infty} \frac{dk}{2\pi} \frac{\sin(kx)\operatorname{sh}(\frac{vk}{2})}{\operatorname{ch}(\frac{k}{2})} = i \int_{0}^{\infty} \frac{dk}{\pi} \frac{\sin(kx)\operatorname{sh}(\frac{vk}{2})}{\operatorname{ch}(\frac{k}{2})}$$

$$= i \int_{0}^{\infty} \frac{dk}{\pi} \frac{\sin(kx) e^{vk/2}}{\operatorname{ch}(\frac{k}{2})} - i \int_{0}^{\infty} \frac{dk}{\pi} \frac{\sin(kx) e^{-vk/2}}{\operatorname{ch}(\frac{k}{2})}$$

the sine-Fourier integral can be evaluated see e.g. [31]

$$= \frac{i}{\pi} \frac{i}{4} \left[\partial_x \ln \Gamma(\tfrac{1-v-2ix}{4}) - \partial_x \ln \Gamma(\tfrac{1-v+2ix}{4}) \right.$$

$$\left. + \partial_x \ln \Gamma(\tfrac{3-v+2ix}{4}) - \partial_x \ln \Gamma(\tfrac{3-v-2ix}{4}) \right]$$

$$- \frac{i}{\pi} \frac{i}{4} \left[\partial_x \ln \Gamma(\tfrac{1+v-2ix}{4}) - \partial_x \ln \Gamma(\tfrac{1+v+2ix}{4}) \right.$$

$$\left. + \partial_x \ln \Gamma(\tfrac{3+v+2ix}{4}) - \partial_x \ln \Gamma(\tfrac{3+v-2ix}{4}) \right]$$

applying the identity $\Gamma(\alpha)\Gamma(1-\alpha) = \pi/\sin(\pi\alpha)$

$$= -\frac{1}{2\pi i}\partial_x \ln\left(\frac{\sin(\frac{\pi}{4} - \frac{\pi}{4}(v+2ix))\sin(\frac{\pi}{4} - \frac{\pi}{4}(v-2ix))}{\sin(\frac{\pi}{4} + \frac{\pi}{4}(v+2ix))\sin(\frac{\pi}{4} + \frac{\pi}{4}(v+2ix))}\right)$$

$$= -\frac{1}{2\pi i}\partial_x \ln\left(\frac{\operatorname{ch}(\pi x) - \sin(\frac{v\pi}{2})}{\operatorname{ch}(\pi x) + \sin(\frac{v\pi}{2})}\right).$$

So that the desired function in position space is

$$\ln y(x) = \operatorname{const} + \ln\left(\frac{\operatorname{ch}(\pi x) - \sin(\frac{v\pi}{2})}{\operatorname{ch}(\pi x) + \sin(\frac{v\pi}{2})}\right) + \ldots; \quad \text{with } v = 1 - (2\chi/i). \quad (E.5)$$

Considering the limit $\chi \to 0$, i.e. $v \to 1$, we find

$$\ln\left(\frac{\operatorname{ch}(\pi x) - \sin(\frac{v\pi}{2})}{\operatorname{ch}(\pi x) + \sin(\frac{v\pi}{2})}\right) = \ln\left(\frac{\operatorname{ch}(\pi x) - \sin(\frac{\pi}{2})}{\operatorname{ch}(\pi x) + \sin(\frac{\pi}{2})}\right)$$

$$= \ln\left(\frac{2\operatorname{sh}^2(\frac{\pi x}{2})}{2\operatorname{ch}^2(\frac{\pi x}{2})}\right)$$

$$= 2\ln\tanh\left|\frac{\pi x}{2}\right|$$

which agrees with the result obtained by starting with principal value integrations.

Appendix F

Contributions from the Lattice to the Eigenvalue

The leading and constant contribution with respect to the system size of the eigenvalue of open XXX chain can be obtained from (4.40) by treating the inhomogeneity s_j as a continuous variable e.g. called z. Explicitly we have by taking the logarithmic derivative

$$\partial_z \ln \Lambda(\tfrac{ic}{2}+z) - \partial_z \ln \Lambda(\tfrac{ic}{2}-z) = \frac{1}{z+\alpha^+} + \frac{1}{z+\alpha^-} + \frac{1}{z-\alpha^+} + \frac{1}{z-\alpha^-}$$
$$+ \frac{1}{z+ic} + \frac{1}{z-ic} - \frac{1}{z+\tfrac{ic}{2}} - \frac{1}{z-\tfrac{ic}{2}}$$
$$+ \sum_{\ell=1}^{L}\left(\frac{1}{z-(ic+s_\ell)} + \frac{1}{z-(ic-s_\ell)}\right.$$
$$\left.+ \frac{1}{z-(-ic+s_\ell)} + \frac{1}{z-(-ic-s_\ell)}\right). \tag{F.1}$$

Applying a complex integration with a closed contour $\{\mathscr{E}\} = -\{\mathscr{E}\}$ invariant under inversion we get for the left hand side

$$2\int_\mathscr{E} \frac{dz}{2\pi i}\partial_z \ln \Lambda(\tfrac{ic}{2}+z) = 2(N-P) \tag{F.2}$$

which counts the zeros (N) and the poles (P) of the desired function $\Lambda(z)$ around the point $\tfrac{ic}{2}$. The pole structure of the right hand side (see figure F.1) reveals for $|\alpha^\pm| > \tfrac{c}{2}$ that the number of zeros and poles of $\Lambda(z)$ is unchanged in the regions shaded in figure F.2. Assuming neither zeros nor poles, i.e. $N = P = 0$, and including the points $\pm\tfrac{ic}{2}$ by the closed contour $\mathscr{E}$ lowers the number $N-P$ by one. This means $\Lambda(z)$ has a first-order pole[1] at $z=0$

---
[1] Alternatively multi-order poles with orders differing by one are possible as well.

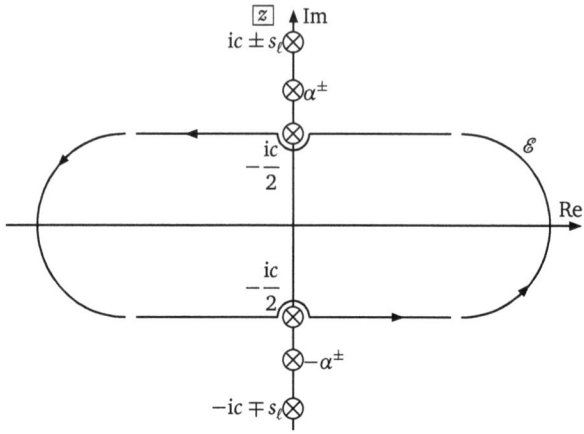

Figure F.1: Zero and pole structure of the right hand side of equation (F.1)

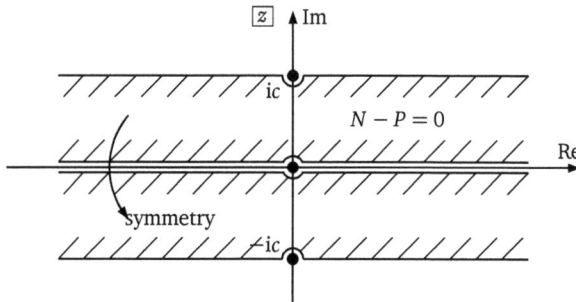

Figure F.2: The shaded regions indicated regions where the number of zeros and poles of $\Lambda(z)$ is unchanged.

or $z = ic$. By choosing a simple pole to be located at $+ic$, and hence also $z = -ic$ due to the symmetry $\Lambda(z) = \Lambda(-z)$, we find $\Lambda(z)$ non-zero and analytic within the strip $|\operatorname{Im} z| < \frac{c}{2}$.

Drawing together the arguments on the left hand side by Fourier Transformation as described in E (here without any χ) we find

$$\mathcal{F}(\partial \ln \Lambda)(k) = -\frac{i\pi}{\operatorname{ch}(\frac{kc}{2})}\left[e^{-k|\alpha^+|} + e^{-k|\alpha^-|} - e^{-\frac{kc}{2}} - e^{-kc} + \sum_{\ell=1}^{L}\left(e^{-k(c+is_\ell)} + e^{-k(c-is_\ell)}\right)\right].$$

Due to the symmetry we restricted $k > 0$ and closed the contours for the right hand side in the lower half plane in order to use Jordan's lemma and made use of the symmetry $\alpha^+ \leftrightarrow -\alpha^+$ of the right hand side. Further we used the fact that $\alpha^\pm$ are purely imaginary. Transforming back we can use the sine Fourier transformation and relate the outcome to digamma functions ψ via

$$\int_0^\infty dk\, \frac{\sin(kx)e^{-ak}}{\operatorname{ch}(bk)} = -\frac{i}{4b}\left[\psi(\frac{a+b+ix}{4b}) - \psi(\frac{a+b-ix}{4b})\right.$$
$$\left. + \psi(\frac{a+3b-ix}{4b}) - \psi(\frac{a+3b+ix}{4b})\right]$$

resulting in

$$\partial_\lambda \ln \Lambda(\tfrac{ic}{2}) = -\frac{i}{c}\left[\psi(|\alpha^+|/2c) - \psi(|\alpha^+|/2c + 1/2) + c/|\alpha^+|\right.$$
$$+ \psi(|\alpha^-|/2c) - \psi(|\alpha^-|/2c + 1/2) + c/|\alpha^-|$$
$$- \psi(1/4) + \psi(3/4) - 2 + \psi(1/2) - \psi(1) + 1$$
$$\left.+ 2L(\psi(1/2) - \psi(1) + 1)\right].$$

For the eigenvalue of the spin chain hamiltonian (4.42) we multiply by ic and evaluate the digamma function where possible

$$ic\frac{\partial \ln \Lambda}{\partial \lambda}(\tfrac{ic}{2}) = \psi(|\alpha^+|/2c) - \psi(|\alpha^+|/2c + 1/2) + c/|\alpha^+|$$
$$+ \psi(|\alpha^-|/2c) - \psi(|\alpha^-|/2c + 1/2) + c/|\alpha^-|$$
$$+ \pi - 2\ln 2 - 1 + (2 - 4\ln 2)L.$$

Bibliography

[1] F. Alcaraz, M. Barber, M. Batchelor, R. Baxter, and G. Quispel, *Surface exponents of the quantum XXZ, Ashkin-Teller, and Potts models*, J. Phys. A **20** (1987), 6397–6409.

[2] F. C. Alcaraz, M. Baake, U. Grimmn, and V. Rittenberg, *Operator content of the XXZ chain*, J. Phys. A **21** (1988), L117–L120.

[3] F. C. Alcaraz, M. N. Barber, and M. T. Batchelor, *Conformal invariance and the spectrum of the XXZ chain*, Phys. Rev. Lett. **58** (1987), 771–774.

[4] T. Ami, M. K. Crawford, R. L. Harlow, Z. R. Wang, D. C. Johnston, Q. Huang, and R. W. Erwin, *Magnetic susceptibility and low-temperature structure of the linear chain cuprate Sr_2CuO_3*, Phys. Rev. B **51** (1995), 5994–6001.

[5] L. Amico, H. Frahm, A. Osterloh, and G. Ribeiro, *Integrable spin-boson models descending from rational six-vertex models*, Nucl. Phys. B **787** [FS] (2007), 283–300.

[6] P. Baseilhac, *New results in the XXZ open spin chain*, Proceedings - RAQIS 2007, 2007.

[7] P. Baseilhac and K. Koizumi, *Exact spectrum of the xxz open spin chain from the q-onsager algebra representation theory*, J. Stat. Mech. (2007), P09006.

[8] M. T. Batchelor, R. J. Baxter, M. J. O'Rourke, and C. M. Yung, *Exact solution and interfacial tension of the six-vertex model with anti-periodic boundary conditions*, J. Phys. A **28** (1995), 2759–2770.

[9] R. J. Baxter, *Exactly solved models in statistical mechanics*, Academic Press, 1982.

[10] V. V. Bazhanov and N. Yu. Reshetikhin, *Critical RSOS models and conformal field theory*, Int. J. Mod. Phys. A **4** (1989), 115–142.

[11] V. V. Bazhanov and V. V. Mangazeev, *Analytic theory of the eight-vertex model*, Nucl. Phys. B **775** (2007), 225–282.

[12] F. Berezin, G. Pokhil, and V. Finkelberg, *The Schrödinger equation for systems of one-dimensional particles with point-like interaction*, Vestnik Moskov. Univ. **1** (1964), 21–28.

[13] H. Bethe, *Zur Theorie der Metalle. I. Eigenwerte und Eigenfunktionen der linearen Atomkette*, Z. Phys. **71** (1931), 205–226.

[14] N. M. Bogoliubov, R. K. Bullough, and J. Timonen, *Exact solution of generalized taviscummings model in quantum optics*, J. Phys. A **29** (1996), 6305–6312.

[15] S. Bose, *Quantum communication through an unmodulated spin chain*, Phys. Rev. Lett. **91** (2003), no. 20, 207901.

[16] E. Boulat, H. Saleur, and P. Schmitteckert, *Twofold advance in the theoretical understanding of far-from-equilibrium properties of interacting nanostructures*, Phys. Rev. Lett. **101** (2008), 140601.

[17] E. Brézin and J. Zinn-Justin, *Un problème à n corps soluble*, C.R. Acad. Sci. Paris Sér. B **263** (1966), 670–673.

[18] J. Cao, H.-Q. Lin, K.-J. Shi, and Y. Wang, *Exact solution of XXZ spin chain with unparallel boundary fields*, Nucl. Phys. B **663** (2003), 487–519.

[19] J. Cao and Y. Wang, *Spin current in quantum XXZ spin chain*, Nucl. Phys. B **792** (2008), 284–299.

[20] R. A. Cowley, D. A. Tennant, S. E. Nagler, and T. Perring, *Spinons and spin waves in one-dimensional heisenberg antiferromagnets*, Journal of Magnetism and Magnetic Materials **140-144** (1995), no. Part 3, 1651 – 1652, International Conference on Magnetism.

[21] J. Damerau, F. Göhmann, N. P. Hasenclever, and A. Klümper, *Density matrices for finite segments of Heisenberg chains of arbitrary length*, J. Phys. A **40** (2007), 4439.

[22] J. de Gier and F. H. L. Essler, *Bethe ansatz solution of the asymmetric exclusion process with open boundaries*, Phys. Rev. Lett. **95** (2005), 240601.

[23] ———, *Exact spectral gaps of the asymmetric exclusion process with open boundaries*, J. Stat. Mech. (2006), no. 12, P12011.

[24] H. J. de Vega and A. González-Ruiz, *Boundary K-matrices for the six vertex and the $n(2n-1)A_{n-1}$ vertex models*, J. Phys. A **26** (1993), L519–L524.

[25] P. Dirac, *On the theory of quantum mechanics*, Proc. Roy. Soc. London A **112** (1926), 661–677.

Bibliography

[26] V. G. Drinfel'd, *Hopf algebras and the quantum Yang-Baxter equation*, Soviet Math. Dokl. **32** (1985), 254–258, translation from Dokl. Akad. Nauk SSSR 283, 1060-1064 (1985).

[27] H.-A. Engel, E. I. Rashba, and B. I. Halperin, *Theory of spin hall effects in semiconductors*, Handbook of Magnetism and Advanced Magnetic Materials (H. Kronmüller and S. Parkin, eds.), John Wiley & Sons Ltd, Chichester, UK, 2007, pp. 2858–2877.

[28] F. H. L. Essler, H. Frahm, F. Göhmann, A. Klümper, and V. E. Korepin, *The One-Dimensional Hubbard Model*, Cambridge University Press, 2005.

[29] F. H. L. Essler, *Sine-gordon low-energy effective theory for copper benzoate*, Phys. Rev. B **59** (1999), no. 22, 14376–14383.

[30] L. D. Faddeev, *Instructive history of the quantum inverse scattering method*, Acta Appl. Math. **39** (1995), 69–84.

[31] L. D. Faddeev and L. A. Takhtajan, *Spectrum and scattering of excitations in the one-dimensional isotropic Heisenberg model*, J. Sov. Math. **24** (1984), 241–267, [Zap. Nauch. Semin. LOMI 109, 134 (1981)].

[32] _____, *Hamiltonian methods in the theory of solitons*, Springer, Berlin, 1987.

[33] H. Frahm and A. Zvyagin, *The open spin chain with impurity: an exact solution*, J. Phys. Condens. Matter **9** (1997), 9939–9946.

[34] H. Frahm, A. Seel, and T. Wirth, *Separation of variables in the open XXX chain*, Nucl. Phys. B **802** (2008), 351–367.

[35] W. Galleas, *Functional relations from the Yang-Baxter algebra: Eigenvalues of the XXZ model with non-diagonal twisted and open boundary conditions*, Nucl. Phys. B **790** (2008), 524–542.

[36] F. Göhmann, A. Klümper, and A. Seel, *Integral representations for correlation functions of the XXZ chain at finite temperature*, J. Phys. A **37** (2004), 7625.

[37] F. Göhmann, *The one-dimensional Hubbard model exakt solution and algebraic structure*, Habilitationsschrift, Universität Bayreuth, 2001.

[38] J. Grelik, *Spinstrom in 1D magnetischen Ketten*, Diplomarbeit, Institut für Theoretische Physik, Leibniz Universität Hannover, 2009.

[39] W. Heisenberg, *Zur Theorie des Ferromagnetismus*, Z. Phys. **49** (1928), 619.

[40] W. Heisenberg, *Mehrkörperproblem und Resonanz in der Quantenmechanik*, Z. Phys. A **38** (1926), 411–426.

[41] K. Hepp and E. H. Lieb, *On the superradiant phase transition for molecules in a quantized radiation field: the dicke maser model*, Ann. Phys. (N.Y.) **76** (1973), 360.

[42] L. Hoddesson, E. Baym, and M. Eckert, *The development of the quantum mechanical electron theory of metals: 1926–1933*, Out of the Crystal Maze: Chapters from The History of Solid State Physics (L. Hoddeson, E. Braun, J. Teichmann, and S. Weart, eds.), Oxford University Press, New York, 1992, pp. 88–181.

[43] Y. Ikhlef, J. L. Jacobsen, and H. Saleur, *A staggered six-vertex model with non-compact continuum limit*, Nucl. Phys. B **789** (2008), 483–524.

[44] E. Ising, *Beitrag zur Theorie des Ferromagnetismus*, Z. Phys. **31** (1925), 263–258.

[45] E. Jaynes and F. Cummings, *Comparison of quantum and semiclassical radiation theory with application to the beam maser*, Proc. IEEE **51** (1963), 89.

[46] M. Jimbo, R. Kedem, T. Kojima, H. Konno, and T. Miwa, *XXZ chain with a boundary*, Nucl. Phys. B **441 [FS]** (1995), 437.

[47] M. Jimbo, T. Miwa, and F. Smirnov, *Hidden grassmann structure in the XXZ model III: Introducing matsubara direction*, J. Phys. A **42** (2009), 304018.

[48] R. Jördens, L. Tarruell, D. Greif, T. Uehlinger, N. Strohmaier, H. Moritz, T. Esslinger, L. D. Leo, C. Kollath, A. Georges, V. Scarola, L. Pollet, E. Burovski, E. Kozik, and M. Troyer, *Quantitative determination of temperature in the approach to magnetic order of ultracold fermions in an optical lattice*, arxiv:0912.3790, to appear in PRL, 2009.

[49] A. Kapustin and S. Skorik, *Surface excitations and surface energy of the antiferromagnetic XXZ chain by the Bethe ansatz approach*, J. Phys. A **29** (1996), 1629–1638.

[50] C. Kim, A. Y. Matsuura, Z.-X. Shen, N. Motoyama, H. Eisaki, S. Uchida, T. Tohyama, and S. Maekawa, *Observation of spin-charge separation in one-dimensional $SrCuO_2$*, Phys. Rev. Lett. **77** (1996), no. 19, 4054–4057.

[51] A. N. Kirillov and N. Yu. Reshetikhin, *Exact solution of the Heisenberg XXZ model of spin s*, J. Sov. Math. **35** (1986), 2627–2643, [Zap. Nauch. Sem. LOMI **145**, 109–133 (1985)].

[52] ———, *The Yangians, Bethe ansatz and combinatorics*, Lett. Math. Phys. **12** (1986), 199.

[53] ———, *Exact solution of the integrable XXZ Heisenberg model with arbitrary spin: II. Thermodynamics*, J. Phys. A **20** (1987), 1586.

[54] N. Kitanine, K. K. Kozlowski, J. M. Maillet, N. A. Slavnov, and V. Terras, *Correlation functions of the open XXZ chain. I*, J. Stat. Mech. (2007), P10009.

[55] _____, *Correlation functions of the open XXZ chain II*, J. Stat. Mech. (2008), P07010.

[56] N. Kitanine, J. M. Maillet, and V. Terras, *Form factors of the XXZ Heisenberg spin-$\frac{1}{2}$ finite chain*, Nucl. Phys. B **554** (1999), 647.

[57] A. Klümper and M. T. Batchelor, *An analytic treatment of finite-size corrections of the spin-1 antiferromagnetic XXZ chain*, J. Phys. A **23** (1990), L189.

[58] A. Klümper, M. T. Batchelor, and P. A. Pearce, *Central charges of the 6- and 19-vertex models with twisted boundary conditions*, J. Phys. A **24** (1991), 3111.

[59] A. Klümper, J. R. R. Martínez, C. Scheeren, and M. Shiroishi, *The spin-1/2 XXZ chain at finite magnetic field: Crossover phenomena driven by temperature*, J. Stat. Phys. **102** (2000), 937.

[60] A. Klümper and P. A. Pearce, *Analytic calculation of scaling dimensions: Tricritical hard squares and critical hard hexagons*, J. Stat. Phys. **64** (1991), 13–76.

[61] _____, *Conformal weights of rsos lattice models and their fusion hierarchies*, Physica A **183** (1992), 304–350.

[62] V. E. Korepin, N. M. Bogoliubov, and A. G. Izergin, *Quantum Inverse Scattering Method and Correlation Functions*, Cambridge University Press, 1993.

[63] P. P. Kulish and N. Y. Reshetikhin, *Quantum linear problem for the sine-Gordon equation and higher representations*, J. Sov. Math. **23** (1983), 2435–2441.

[64] P. P. Kulish, N. Yu. Reshetikhin, and E. K. Sklyanin, *Yang-Baxter equation and representation theory: I*, Lett. Math. Phys. **5** (1981), 393–403.

[65] P. P. Kulish and E. K. Sklyanin, *Quantum spectral transform method. Recent developments.*, Integrable Quantum Field Theories (Berlin) (J. Hietarinta and C. Montonen, eds.), Lecture Notes in Physics, vol. 151, Springer Verlag, 1982, pp. 61–119.

[66] A. Kuniba, T. Nakanishi, and J. Suzuki, *Functional relations in solvable lattice models: I. Functional relations and representation theory*, Int. J. Mod. Phys. A **9** (1994), 5215–5266.

[67] J. B. McGuire, *Interacting fermions in one dimension. I. Repulsive potential*, J. Math. Phys. **6** (1965), 432–439.

[68] C. S. Melo, G. A. P. Ribeiro, and M. J. Martins, *Bethe ansatz for the XXX-S chain with non-diagonal open boundaries*, Nucl. Phys. B **711** (2005), 565–603.

[69] L. Mezincescu and R. I. Nepomechie, *Fusion procedure for open chains*, J. Phys. A **25** (1992), 2533–2543.

[70] T. Muir, *A Treatise on the Theory of Determinants by Thomas Muir. Revised and enlarged by William H. Metzler*, Dover, New York, NY, USA, 1960, Corrected printing of the 1933 edition.

[71] R. Murgan and R. I. Nepomechie, *Addendum to 'Bethe ansatz derived from the functional relations of the open XXZ chain for new special cases'*, J. Stat. Mech. (2005), P11004.

[72] R. Murgan, R. I. Nepomechie, and C. Shi, *Boundary energy of the open XXZ chain from new exact solutions*, Ann. H. Poincaré **7** (2006), 1429–1448.

[73] _____, *Exact solution of the open XXZ chain with general integrable boundary terms at roots of unity*, J. Stat. Mech. (2006), P08006.

[74] R. I. Nepomechie, *Solving the open XXZ spin chain with nondiagonal boundary terms at roots of unity*, Nucl. Phys. B **622** (2002), 615–632.

[75] _____, *Functional relations and Bethe ansatz for the XXZ chain*, J. Stat. Phys. **111** (2003), 1363–1376.

[76] _____, *Bethe ansatz solution of the open XXZ chain with nondiagonal boundary terms*, J. Phys. A **37** (2004), 433–440.

[77] S. Niekamp, *Die XXZ-Spinkette mit nichttrivialen Randbedingungen*, Diplomarbeit, Institut für Theoretische Physik, Leibniz Universität Hannover, 2008.

[78] S. Niekamp, T. Wirth, and H. Frahm, *The XXZ model with anti-periodic twisted boundary conditions*, J. Phys. A **42** (2009), 195008.

[79] M. Niss, *History of the Lenz-Ising model 1920–1950: From ferromagnetic to cooperative phenomena*, Arch. Hist. Exact Sci. **59** (2005), 267–318.

[80] P. A. Pearce and A. Klümper, *Finite-size corrections and scaling dimensions of solvable lattice models: An analytic method*, Phys. Rev. Lett. **66** (1991), 974.

[81] J. Plefka, *Spinning strings and integrable spin chains in the AdS/CFT correspondence*, LivingRev.Rel. **8** (2005), no. 9.

[82] A. Seel and T. Wirth, *Non-linear integral equations and determinant formulae of the open XXZ spin chain*, J. Phys. A **42** (2009), 115202, arXiv:0808.2108.

[83] E. K. Sklyanin, *Boundary conditions for integrable quantum systems*, J. Phys. A **21** (1988), 2375–2389.

Bibliography

[84] _____, *Quantum inverse scattering method. Selected topics*, Quantum Group and Quantum Integrable Systems (M.-L. Ge, ed.), Nankai Lectures in Mathematical Physics, World Scientific, Singapore, 1992, pp. 63–97.

[85] S. Skorik and H. Saleur, *Boundary bound states and boundary bootstrap in the sine-gordon model with dirichlet boundary conditions*, J. Phys. A **28** (1995), 6605–6622.

[86] M. Takahashi, *Thermodynamics of the Heisenberg-Ising model for $|\Delta| < 1$ in one dimension*, Phys. Lett. **36A** (1971), no. 4, 325–326.

[87] _____, *Thermodynamics of One-Dimensional Solvable Models*, Cambridge University Press, 1999.

[88] _____, *Simplification of thermodynamic Bethe-ansatz equations*, Physics and Combinatorics – Proceedings of the Nagoya 2000 International Workshop (A. N. Kirillov and N. Liskova, eds.), World Scientific, 2000, Nagoya University, 21 – 26 August 2000, pp. 299–304.

[89] M. Takahashi, M. Shiroishi, and A. Klumper, *Equivalence of TBA and QTM*, J. Phys. A **34** (2001), L187–L194.

[90] M. Tavis and F. W. Cummings, *Approximate solutions for an N-molecule-radiation-field hamiltonian*, Phys. Rev. **188** (1969), 692.

[91] _____, *Exact solution for an N-molecule-radiation-field hamiltonian*, Phys. Rev. **170** (1969), 379.

[92] D. A. Tennant, T. G. Perring, R. A. Cowley, and S. E. Nagler, *Unbound spinons in the S=1/2 antiferromagnetic chain $KCuF_3$*, Phys. Rev. Lett. **70** (1993), no. 25, 4003–4006.

[93] J. Vermaseren, *New features of FORM*, arXiv:math-ph/0010025, 2000.

[94] A. Wallraff, D. I. Schuster, A. Blais, J. M. Gambetta, J. Schreier, L. Frunzio, M. H. Devoret, S. M. Girvin, and R. J. Schoelkopf, *Sideband transitions and two-tone spectroscopy of a superconducting qubit strongly coupled to an on-chip cavity*, Phys. Rev. Lett. **99** (2007), 050501.

[95] Y.-S. Wang, *The reconstruction of local quantum operators for the boundary XXZ spin-1/2 Heisenberg chain*, J. Phys. A **33** (2000), 4009–4014.

[96] _____, *The scalar products and the norm of Bethe eigenstates for the boundary XXX Heisenberg spin-1/2 finite chain*, Nucl. Phys. B **622** [**FS**] (2002), 633.

[97] T. Wirth, *Spektrum der XXZ-Spinkette mit Randfeldern*, Diploma thesis, Institut für Theoretische Physik, Universität Hannover, 2006.

[98] C. N. Yang, *Some exact results for the many-body problem in one dimension with repulsive delta-function interaction*, Phys. Rev. Lett. **19** (1967), 1312–1315.

[99] W.-L. Yang, R. I. Nepomechie, and Y.-Z. Zhang, *Q-operator and T-Q relation from the fusion hierarchy*, Phys. Lett. B **633** (2006), 664–670.

[100] C. M. Yung and M. T. Batchelor, *Integrable vertex and loop models on the square lattice with open boundaries via reflection matrices*, Nucl. Phys. B **435** (1995), 430–462.

[101] Y.-k. Zhou, *Fusion hierarchy and finite-size corrections of $U_q[sl(2)]$-invariant vertex models with open boundaries*, Nucl. Phys. B **453** (1995), 619–646.

[102] _____, *Row transfer matrix functional relations for baxter's eight-vertex and six-vertex models with open boundaries via more general reflection matrices*, Nucl. Phys. B **458** (1996), 504–532.

[103] Y.-K. Zhou and P. A. Pearce, *Solution of functional equations of restricted $A_{n-1}^{(1)}$ fused lattice models*, Nucl. Phys. B **446** (1995), 485–510.

Publications

The subsequent list provides the publictations which emerged during the work on this thesis.

[A] A. Seel and T. Wirth, *Non-linear integral equations and determinant formulae of the open XXZ spin chain*, J. Phys. A **42** (2009), 115202, arXiv:0808.2108

[B] H. Frahm, A. Seel, and T. Wirth, *Separation of variables in the open XXX chain*, Nucl. Phys. B **802** (2008), 351–367, arXiv:0803.1776

[C] S. Niekamp, T. Wirth, and H. Frahm, *The XXZ model with anti-periodic twisted boundary conditions*, J. Phys. A **42** (2009), 195008, arXiv:0902.1079

[D] Luigi Amico, Holger Frahm, Andreas Osterloh, and Tobias Wirth, *Separation of variables for integrable spin-boson models*, Nucl. Phys. B **839** (2010), 604-626, arXiv:1005.5681

[E] Holger Frahm, Jan H. Grelik, Alexander Seel, and Tobias Wirth, *Functional Bethe ansatz methods for the open XXX chain*, J. Phys. A **44** (2011), 015001, arXiv:1009.1081

Acknowledgements

I am indebted to a lot of people without whom this thesis would never have been possible to complete.

Sincere thanks are given to Prof. Dr. Holger Frahm who accepted me as his diploma student and gave me the opportunity and trust for my PhD studies. His advice and guidance was always very helpful and more than welcome and I am very grateful having had him as my supervisor.

Similarly great impact had Dr. Alexander Seel on me and my studies. I want to thank him not only for working with me on several projects and reviewing the manuscript but also for the great times we had together as office mates during the last years.

The great atmosphere in our group was a big help at all times and I want to thank all members who contributed: André Grabinski, Carsten von Zobeltitz, Jan Grelik, Jörn Bröer, Dr. Peter Finch, Dr. Guillaume Palacios and Sönke Niekamp.

Further I want to thank Prof. Dr. Luis Santos for kindly agreeing to review this thesis. And again I need to thank André Grabinski for extensively going through the drafts of the thesis and helping me with the phrasing.

Prof. Dr. Hans-Jürgen Mikeska influenced me a lot since starting at the university and I thank him for the encouragement he gave me. I also want to thank Priv. Doz. Dr. Andreas Osterloh and Dr. Luigi Amico for the opportunity to participate in the Vigoni project and the collaboration on the spin-boson model. The stay at the DMFCI in Catania was a very pleasant and enjoyable experience.

I could never have finished my studies without having great support in my private life. First of all I want to express my deepest thanks to Renana Iwannek for the love, patience and sympathy she gives to me at all times. My special girls Gwen and Naemi cheer me up every day and Nelly will be kept in our hearts forever.

My parents Agnes and Ewald were the ones who always gave me the opportunity to pursue the goals and dreams I had and I am very grateful. My sister Kerstin had an open ear to my problems and I thank her for all advices and hints she had for me.

Finally I want to thank our secretaries Emma Schwebs and Gitta Richter for the help with the administrative business. This work was supported by the Deutsche Forschungsgemeinschaft which I am really thankful for.

I want morebooks!

Buy your books fast and straightforward online - at one of world's fastest growing online book stores! Environmentally sound due to Print-on-Demand technologies.

Buy your books online at
www.morebooks.shop

Kaufen Sie Ihre Bücher schnell und unkompliziert online – auf einer der am schnellsten wachsenden Buchhandelsplattformen weltweit! Dank Print-On-Demand umwelt- und ressourcenschonend produziert.

Bücher schneller online kaufen
www.morebooks.shop

KS OmniScriptum Publishing
Brivibas gatve 197
LV-1039 Riga, Latvia
Telefax: +371 686 204 55

info@omniscriptum.com
www.omniscriptum.com

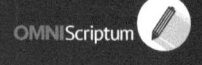

Printed by Books on Demand GmbH, Norderstedt / Germany